Nuts & Bolts

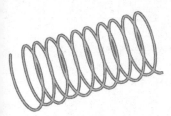

Nuts & Bolts

Seven

Small Inventions

That Changed the World

(in a Big Way)

ROMA AGRAWAL

W. W. NORTON & COMPANY

Celebrating a Century of Independent Publishing

For information about permission to reproduce selections from this book, write to
Permissions, W. W. Norton & Company, Inc., 500 Fifth Avenue, New York, NY 10110

For information about special discounts for bulk purchases, please contact
W. W. Norton Special Sales at specialsales@wwnorton.com or 800-233-4830

Image on page 89: Spring mount drawn by Norm Mason, used
with permission from Mason Industries Incorporated.
Image on page 134: Facsimile of the Leeuwenhoek microscope in Utrecht
University. Wellcome Collection. 4.0 International (CC BY 4.0).
Image on page 150: 1960 of lens development timeline. Used with permission from
The Focal Encyclopedia of Photography, desk edition, London: Focal Press.
Image on page 171: Tampuri. ShareAlike 3.0 Unported (CC BY-SA 3.0).
Image on page 181: Invention of a water pump, miniature from the Book of
Knowledge of Ingenious Mechanical Devices by Al-Jazari, 1203, Turkey. Istan-
bul, Topkapi Sarayi Muzesi Kutuphanesi (Library) © De Agostini Editore.

Manufacturing by Lakeside Book Company
Book design by Chris Welch
Production manager: Erin Reilly

ISBN: 978-1-324-02152-0

W. W. Norton & Company, Inc., 500 Fifth Avenue, New York, N.Y. 10110
www.wwnorton.com

W. W. Norton & Company Ltd., 15 Carlisle Street, London W1D 3BS

1 2 3 4 5 6 7 8 9 0

And finally,
for you, my Flirtman.

CONTENTS

INTRODUCTION

Numerous severed crayons lay in disarray before me. I sighed. The results were disappointing.

I must have been about five years old, and was living with my parents and sister in snowy upstate New York. It was the 1980s, and I owned a selection of large, rectangular lunch boxes that normally housed some sandwiches, snacks, and a thermos. The one open in front of me was my favorite, adorned with a picture of the Muppets on the front. It was not filled with food, though. Instead, it held my huge collection of crayons: long, short, thick, thin, in every shade available. Like most children, I was continuously curious, and one day decided to "discover" what was inside my crayons. So, I peeled off the paper that enveloped them, then held them one at a time against the sharp edge of the open box and snapped them in two. My great anticipation was rather dampened to find, well, just more crayon inside. Nevertheless—and much to the dismay of my sister—I persisted.

When I was a little older and started writing words on paper with pencils, I would twist the pencils inside a sharpener, creating long spirals of shavings to see if the gray rod that marked my sheets went all the way through the body of the pencil. It did. From there, I graduated to pens, dismantling these to reveal rather more exciting interiors. Far from the disappointing crayons of my early childhood, the insides of fountain pens and ballpoints contained slender cartridges and helical springs, held together with a top that threaded, screwlike, onto the rest of the pen.

In addition to taking things apart myself to satisfy my curiosity, I poked my nose in when others did it, too. Growing up in India, I saw my television taken apart when the picture ended up with black lines across it. It was dismantled to reveal boggling innards that I only made sense of when I did a degree in physics. In fact, the reason I chose to study physics was because I wanted to understand the building blocks of our universe. Toward the end of my school years, I had become fascinated by atomic and particle physics, enthralled by the idea that the atom itself, once thought indivisible, was then revealed to be made up of electrons, protons, and neutrons—and that these, after having their turn on the podium of "fundamental building blocks of matter," were supplanted by their even smaller constituents, quarks. Whether or not I understood it at the time, I was on a mission to understand what things were made from, and how they came to be.

Whether it's the matter that makes our universe, living biological creatures, or the human-made objects we invented, complex compositions are made up of smaller and simpler, well, things. I have been lucky enough to carry my childhood curiosity about what makes up objects with me into my career. As an engineer, I am endlessly fascinated by how our machines, buildings, and everyday objects came to be, and what lies at their heart—a fascination that, no doubt, many of you share. This book is the manifestation of that fascination.

Engineering is a vast discipline, but some of its mightiest achievements have been small in scale. Inside all the human-made things around us are fundamental building blocks without which our complex machinery wouldn't exist. At first glance, they might seem uninteresting. Often small, and sometimes hidden, the truth is that each of these elements is an extraordinary feat of engineering with fascinating stories that go back hundreds, if not thousands, of years. During the Renaissance, scientists and engineers defined six "simple machines," described as being the basis of all complex machines. These were the lever, the wheel and axle, the pulley, the inclined plane, the wedge, and the screw. But today, those six feel outdated and insufficient. So, I got rid of a few and added some others to showcase seven elements that I believe form the basis of the modern world. They encompass a vast range of innovations in terms of their underlying scientific principles, the fields of engineering they touch, and the scale of objects they have enabled.

Each of the seven objects I've selected—the nail, the wheel, the spring, the magnet, the lens, the string, and the pump—are wonders of design that went through many different iterations and forms, and continue to do so. They have endured ever since their first appearance. As they evolved, as they were combined in different permutations, the complexity of the machines we could make escalated in a cascading butterfly effect of invention and innovation. Every single one of these objects has touched us individually, and made an indelible mark on our world; without them, our lives would be unrecognizable. They have created and changed our technology, of course, but have also had a sweeping impact on our history, society, political and power structures, biology, communication, transportation, arts, and culture.

I selected these seven objects during the first 2020 lockdown in England, in the midst of a global pandemic. Trapped at home, I let my mind roam free, looking around at my possessions and the objects I could see from my window, and mentally (or sometimes

physically) deconstructing them to see what lay inside. I revisited
the ballpoint pen and saw it built around a spring, a screw, and a
revolving sphere. The blender I used to make my baby's food relied
on gears, which in turn couldn't exist without the wheel. Before
that, when I was breastfeeding, a breast pump allowed my husband to feed our daughter, too. The process of IVF I went through
in order to make the embryo that grew into her, and the research
on creating Covid-19 vaccines, relied on a lens to see things on a cel-
lular scale. The protective masks we wore during our short walks,
and that kept medics safe, were formed of countless fibers twisted
together to form fabric. The speaker on my phone through which I
could hear the voices of my family and friends relied on a magnet,
as did the Ethernet socket that gave me access to the internet.

Even as I thought about larger and more complex objects—
diggers, skyscrapers, factories, tunnels, electrical grids, cars,
satellites, and so on—again and again, I came back to the same
seven foundational innovations. We join things together: the nail.
We need something that rotates or revolves: the wheel. We need
power, and technology that can store it: batteries, sure, but more
fundamentally, the spring. Magnetism (and electricity) allows us
to manipulate things from a distance; the lens lets us play with the
path of light. String gives us a strong material that is also flexible.
To move water and keep us alive, we fashion pumps.

The invention or discovery of each of these seven pieces of engineering involved a process of failure and iteration: of having a need,
then trying out different materials, shapes, and forms, until something worked. To take one example, buildings, bridges, factories,
tractors, cars, phones, locks, watches, and washing machines—in
fact, most things that need pieces of metal to be attached to each
other—have nails, screws, rivets, and bolts keeping them together.
The nail was originally used to join pieces of wood together: a new
concept to create more robust ships and furniture. Later, the screw

vastly improved on the nail's holding power, although it was much harder to make. Then, when thin metal sheets could be cheaply manufactured, neither the nail nor the screw was fit for the purpose, and the rivet came into being. Small rivets in cooking vessels gave way to larger and stronger rivets to join metal planes, ships, and bridges, before engineers invented the bolt, a combination of the rivet and screw, which was stronger and easier to install. The Shard, the tallest tower in Western Europe and a project that I worked on for six years as a structural engineer, is held stable and robust by such bolts.

All this evolution doesn't mean that the original nail is obsolete, however. In fact, nails, and their multiple reincarnations, are all being used in parallel with screws, rivets, and bolts, each one for the purpose it best suits. And that's how design changes: sometimes we use the same technology for centuries before we suddenly invent a new material or process, and realize that we need to adapt existing technology to suit. Other times, it's the other way around: we invent a new technology, like the incredibly strong fiber Kevlar, and then find purposes for it—in this case, bulletproof vests. Some of these inventions developed independently in different parts of the globe with very similar designs, like the wheel, but others, like the pump, looked very different. And so, these inventions were born, then changed and evolved in their own ways, often going on to have unexpected applications and implications far beyond their original purpose.

While we think of engineering as a field littered with inanimate objects and complex pieces of technology that often feel alien or beyond our understanding, at the heart of engineering is people: those who create it, those who need and use it, those who sometimes inadvertently make a contribution to it. They are the seamstress in Delaware worrying about Neil Armstrong's gusset holding, the doctor who passed electrical current through his hands, the shop owner who studied his sperm under a microscope, and the sickly recipient of a pig's heart. They are the Indian polymath who directed radio waves through the body of an important governor, the immigrant chemist who thought she'd made an error but invented something incredible, the Islamic scholar who changed the way we see, and the housewife who got frustrated with broken dishes. For centuries, engineering has been dominated, particularly in the West, by the rich and educated—and, historically, by men. The stories, devices, and innovators I've chosen to feature here are from all over the world, from different eras, and include the often hidden or unacknowledged contribution of minoritized people in engineering: stories that are often lost because their work wasn't documented, they didn't (or couldn't) apply for patents, or weren't granted them.

In the pages that follow, I will show you that engineering is the meeting of science, design, and history. It's about human need and creativity, about finding problems and creating solutions to them in ways that haven't been attempted before. It's about trying to make our lives better, but knowing that, conversely, our inventions can have a devastating impact on society when not used responsibly. I will show you how engineering at its most fundamental is inextricably linked to your everyday life, and to humanity. My hope is that I will reignite your childhood curiosity and inspire you to investigate the increasingly complicated black box of engineering, in order to understand the building blocks of our world a little bit better.

Nail

"Red hot? Not hot enough," yells Rich over the din of the workshop.

With my bare hands, I've been gingerly holding on to the top of a thin steel rod about the length of my arm. The bottom is submerged in coke burning in a brick furnace at over 1,000°C. An electric fan blows air on the flame to make it even hotter. But when I pull the rod out to check its color, which is an angry glowing red, Rich, the blacksmith, tells me it's not ready yet. So I shove it back into the fire until it brightens to a searing orangey-yellow. Now, with this luminous piece of steel, I can get to work on making a nail.

For this, I need tools. Before me is an anvil—the classic hulking iron blacksmith's block—with the legend "102 kgs" stamped on its side. Angling the rod so its glowing end rests on the flat surface of the anvil, I hit the end with a heavy hammer, and the metal flattens a little. I hit it a few more times, then rotate the rod by ninety degrees, and hit it again. After a while, though, the end of the rod darkens; I can feel the hammer bounce, and the sound of the clanging becomes sharper. The rod has cooled. So back in the fire it goes until it reaches orangey-yellowness again. Hammer, rotate, reheat. It takes me three cycles of this to fashion a fairly good taper at the end of the rod. (Rich can do it in one.) But finally, I have the body of my nail.

Next, I slot a chisel—sharp end pointing up—into a square hole in the anvil known as the "hardy hole." I lay my reheated, tapered rod across the chisel blade and hit it hard (as hard as nails, perhaps) to notch two sides, which will make it easier to separate the tapered end from the rest of the rod. The last tool I need is the heading tool: a long, rectangular metal plate studded with different-sized holes, as though it's been attacked by a hole punch. Choosing a hole that is large enough to allow most of the taper through, I insert my nail-to-be. With some strenuous twisting, I break the rest of the rod away from the nail. Now the point of the nail hangs down through the hole, with about a centimeter of head poking through the top. After lining up the point of the nail with a circular hole on the anvil, I quickly bash the top to flatten it and create the nail head.

Finally, the "quenching": I plunge the whole plate and nail into a trough of water, producing a satisfying sizzle and clouds of steam as hot hits cold. I pull them out, give a gentle tap with the hammer to release the nail from the heading tool, and it falls to the floor with a clang. My humble, and still slightly warm, creation.

Nails may seem like a simple piece of engineering, but take a look around you, and you'll see that they are everywhere. From my desk, as I write, I can see pictures suspended from them and bookshelves fixed together with them. The desk itself is secured with them, as are the shoes I've just kicked off underneath. The walls, made from panels of wood underneath the plaster, and the floor, resting on wooden joists, are all joined with nails. Most of these nails I can't actually see, because their bodies are buried in wood, leather, and brick, but I'm aware of their silent, reassuring presence.

The nail enables us to connect things together. That might not sound like much, but the act of joining two things was once radical. Almost everything around us that is human-made is essentially a coming together of different segments and materials, a fact that we take for granted. But it wasn't always the case. Many millennia ago, creating something usually meant shaping a single piece of material into a more purposeful form: cutting a chamber into a rock to create a cave, for example, or sharpening a stone to make a tool, or toppling a tree trunk over a stream to form a bridge. These are all useful pieces of engineering. However, to build more complex dwellings, or a weapon with a stone point attached to a pole, or a bridge to span a distance that a single log cannot, we needed to be able to fasten segments together—allowing a big step up in the complexity of our creations.

Of course, you can always stack up rocks to support a bridge. You can tie things with ropes and leather, or, once glue had been invented, stick them together. But nails, and the derivatives they spawned—the rivet, the screw, the bolt—enabled robust connections between different materials to be created by anyone, with little guidance, and at a vast range of scales: we could join large timber beams and columns together to make buildings, connect layers of wooden planks to build boats, and overlay thin sheets of metal to make ships, sculptures, locks, and watches. Imagine a world without nails. We could never send intricately assembled satellites into space bound only by string, or make watches with moving parts that are glued.

I made my nail at the forge in the village of Much Hadham in Hertfordshire. The forge has been in continuous use since 1811.

My attempt was square-shafted and rather thick and uneven, and creating it was hard work—all that hammering gave me a couple of blisters on my palm, and a trembling bicep. Now, of course, most nails are made by machine, but for thousands of years, from the Egyptians and Romans onward, nail-making essentially involved the processes and the pounding that I'd had to do at Much Hadham.

That hard labor was largely down to the malleability of material used. The story of the nail is also, up to a (tapering) point, the story of metal. While there are sometimes benefits to making nails from other materials, like wood, metal nails were the game changers. Metal has two properties that, together, make it invaluable for nail-making. First, it's strong enough to be hammered straight into other materials to make a connection. But, equally importantly, it can be shaped into a sharp point: metals have a crystalline internal structure, which gives them a special type of flexibility—known as ductility—as the crystals can move slightly over each other. This is why you can open out a paper clip or deform it repeatedly without it breaking. (At the other end of the spectrum are brittle materials like glass, which easily shatter when forced.)

Heat is one way of making metal ductile. The high temperature of a forge excites the floating electrons and the atoms within the crystalline structure, so they move around vigorously. This means heat energy moves through metals quickly, making them good conductors. Different metals have different melting points and different levels of conductivity, but the hotter the metal, the more the atoms and electrons slip and slide over each other, making the material soft and pliable and ready to be hammered into shape. Amazingly, heating and hammering also change the actual structure of the metal, rearranging the large, coarse crystals into smaller, more regular ones, making it stronger, harder, and consistent when it cools.

Humans began fashioning gold during the Stone Age, around

8,000 years ago, and then discovered copper, silver, and lead. Most of these metals are far too soft for making nails, but copper was the first metal that showed potential. Then, some of our most enterprising ancestors figured out how to mix copper and tin to make bronze, finally creating a material that was strong enough to form more durable tools, weapons, armor, and nails.

The oldest bronze nails date back to 3400 BCE and were found in Egypt. They look remarkably like the hand-forged nails made from steel today, although the intervening 5,000 years have left them encrusted, blunt, and discolored. The Egyptians were proficient in working bronze, creating intricate inlays in the metal with precious stones, enamels, and gold, alongside using it for more practical tools like nails, with which they used to put together their boats and chariots.

For a thousand years thereafter, we used copper and bronze to make nails, although bronze was never the most practical commercial choice of material, as it relied on copper and tin, metals that are rarely found in the same place. Around 1300 BCE, metal workers in India and Sri Lanka discovered how to make iron (which is about as hard as bronze, and harder than copper). This ushered in the Iron Age in the East, and bronze was soon supplanted by the new material. The final nail in the coffin for bronze was a politically tumultuous period in the Middle East that started in about 1200 BCE, disrupting trade routes and making tin (and therefore bronze) very expensive. Later, iron too was supplanted, once people discovered that mixing it with a little bit of carbon created alloys like steel, which led to much stronger nails.

The Romans became very skilled in manipulating Indian iron, using it in a variety of ways, from making armor to mass-manufacturing nails across their empire, including Britain. In the 1960s, a huge hoard of Roman nails was found in a field in Perthshire, Scotland, at the site of a Roman legionary fortress, called

Inchtuthil. About 5,000 soldiers of the twentieth legion were based there from around 83 to 86 CE, when it was suddenly abandoned. It was occupied so briefly that the Romans hadn't finished building their usual baths and aqueducts to supply water, but an archaeological dig nevertheless offered fascinating insights into Roman fortress design and their manufacturing capabilities.

The fortress was immense, covering around fifty-three acres (that's over twenty-six soccer fields), and included sixty-four barrack blocks, a hospital, granaries, and, importantly, a fabrica or forge. A smithing hearth was found in one wing of this building, and in another, a large sealed pit. When the archaeologists dug carefully through 2 meters of gravel, they found unexpected treasure: iron nails—875,428 of them, to be exact—in a range of sizes. Remarkably, most were in almost pristine condition. The outermost nails had corroded to form an impermeable crust, protecting the rest from rust, leaving these two-millennia-old nails almost as sharp and shiny as the day they were made.

Outside the fabrica, well-defined ruts had been etched into the streets, indicating that heavy materials had been brought back and forth. The Inchtuthil forge may have supplied nails to other settlements, and the size of the hoard suggests that the Roman governor of Britain at the time, General Julius Agricola, may have planned to build more forts farther north before being suddenly summoned back to Rome as military planning shifted focus toward Europe. When his legion withdrew, they burned down the fortress and buried ten tonnes of nails, fearing the local Caledonians might otherwise melt them down for weapons.

Six distinct types of nails were found in the Inchtuthil pit, each seemingly designed for a different purpose. The most common are small, disc-headed nails, up to the length of a finger and probably used in furniture, or wall and floor paneling. Several types of much larger nails, about the length from my elbow to my fin-

gertips, would have secured heavy timbers, and were designed
with pyramidal heads that could withstand prolonged hammering.
Most of the nails were square-shafted, but among the other types
were twenty-eight round-shanked nails with flattened cone-shaped
heads and chisel-shaped points—these were probably used for pen-
etrating masonry, since a square shaft was in danger of splitting
stone at its corners.

The Romans had hit the nails on their heads: these were
high quality, and very consistent in shape, size, and material.
The larger ones had more carbon than the smaller, making
them harder—which suggests the smiths graded the raw material before forging. Their tips were harder than the heads, probably because of the way they had been heated, hammered, and
quenched. Clearly, they were the work of extremely skilled metal-
workers. As I now know only too well, making nails by hand is
a complex—and exhausting—business. You need to understand
the science of getting the metal to heat up to the ideal temperature, and to use the correct force and direction of hammering—
and this all needs to be done quickly, while the metal is still hot
enough. The ancients couldn't measure what temperature their
materials were reaching, so they were guided by color. Red hot
(around 700–900°C for steel) is fine for bending, but if you try
anything more complex, the metal might crack. It becomes more
pliable when hotter still, turning orange like the evening sun.
Heat it further, to over 1,300°C, and steel glows a blinding white.
This is a good temperature for hammering pieces together to "fire
weld" them, a process that gets its name from the brilliant white
sparks that fly off the steel. (When it's this hot, metal has its
own particular hypnotic intensity. During research for this book,
I spoke to the blacksmith and artist Agnes Jones, who creates
extraordinary organic sculptures in steel. For Agnes, white-hot
steel has a surreal beauty because a thin layer of the surface

melts, blurring its solid lines, like the top layer of a sandy beach on a windy day.) That blinding white heat is as far as a blacksmith wants to go. Take it beyond this point, and steel turns into a sparkler and smells like fireworks. That means it is burning.

So, 1,000–1,200°C is the sweet spot for forging low-carbon steel (the exact temperature will depend on the specific composition of the metal). This is the moment when the metal glows sun-on-a-summer-afternoon yellow, showing that the steel has become soft enough to be manipulated to a point. Once you're happy with the shape, cooling the rod suddenly by plunging it into cold water—the quenching process I followed earlier—helps to harden it, giving it strength and maintaining its shape.

Nail-making continued to be a prized and special skill for centuries in Europe after the Roman Empire fell. In medieval Britain, nailers (the origin of the surname Naylor) created nails for horseshoes, joinery, and house construction. It's difficult to imagine now, but nails were so valued in this preindustrial era, where materials and skilled workers weren't readily available, that the British banned their export to their colonies, including North America, where timber housing was the norm. As a result, nails became so precious there that some people even set fire to their homes when moving in order to retrieve the nails from the ashes. In 1619, a law was passed in the state of Virginia to discourage this practice by promising the owner compensation:

> That it shall not be lawfull for any person so deserting his plantation as aforesaid to burne any necessary houseing that are scituated therevpon, but shall receive so many nailes as may be computed by 2 indifferent men were expended about the building thereof for full satisfaction . . .

After Independence in 1776, the Americans sought to establish their own nail-making industry to serve an expanding economy and housing market. One of the earliest large operations was set up by founding father Thomas Jefferson. Seven years before he became president in 1801, Jefferson started a foundry at his Monticello farm in Charlottesville, Virginia. The mansion, with its 5,000 acres of plantation, was situated at the top of a steep hill, where over 400 enslaved individuals worked during Jefferson's lifetime. One, Joe Fosset, toiled in the nailery from the time he was twelve, alongside other young boys, who together made between 8,000 and 10,000 nails every day by hand—enough to fund the Jefferson family while the depleted soils of the plantation replenished themselves during fallow years. Fosset later became an enslaved foreman in the nailery, and set up his own nailery business after he was freed in order to buy the freedom of his wife and ten children.

Jefferson was proud of his nail-making business, writing to French politician Jean Nicolas DeMeunier that being a nail-maker was, to him, like a "title of nobility." A few years later, when the price of iron had increased and nails were becoming more readily available because the British finally began exporting, Jefferson wrote excitedly to another friend in 1796 of his hopes of increasing production with a new "cutting machine."

This machine, purchased by Jefferson from a Mr. Burral in New York, represented the beginnings of mechanized nail-production. Although nail-making machines had appeared as early as 1600, they had not gained much popularity; they were clumsy to operate and only made one nail at a time. Jefferson's machine cut small, four-penny nails—so-called because a hundred had cost four pennies in medieval Britain—from thin strips of iron used as barrel hoops, probably using a pair of vertical blades operated by a shaft that was turned manually. Jefferson's machine doesn't seem to have been able to create heads for the nails, but other machines from that period used a series of levers to press the wider end of the nail and flatten it. While this cut down the hard physical labor involved in nail-making and sped up the process, it didn't represent a radical break with the past. Square-cut and somewhat chunky, these early machine-made nails had more in common with their handmade predecessors than the perfectly rounded nails of today.

Britain remained a large producer of nails throughout the nineteenth century. A skilled nailer could make a nail by hand in about one minute or less. It was common for boys and girls aged seven, or even younger, to make nails in order to earn every penny they could. Nail-making thrived, particularly in the Black Country in central England, because of the local availability of iron and coal. It was a task typically left to women to earn money for food and household expenses, and so the local colliers, as coal miners were known, and ironmongers often aspired to marry a "nailing wench" who could generate some family income on the side, when they weren't running their homes and looking after their children.

As so often happens with women earning money, however, the practice attracted opposition, with the nail-forgers' union trying to introduce restrictions on women making nails; this was, in turn, opposed by the nail masters, who used them as cheap labor. Some women defied the odds to become successful in their own right. Eliza Tinsley, for instance, became known as "The Widow" when she took over her late husband's nail-making firm after he died in 1851. Tinsley had five surviving children under the age of eleven at the time of her husband's death, and she succeeded in expanding the family's nail- and chain-making business so that it was the largest of its kind in the county of Staffordshire, with warehouses in seven other locations. She was known as a fair and humane employer and traveled across the UK to visit her customers. By the time she died in 1882, aged sixty-nine, her company employed more than 4,000 people. The company exists to this day, still bearing her name, as does each packet of nails it produces.

Overlapping with Eliza's lifetime, however, the industry was revolutionized by two developments. First, engineers discovered the power of precision and mass manufacture: the ability to make things consistently to exact measurements, repeatedly. Henry Maudslay (born in 1771) invented the first practical metal-cutting lathe. Before this, small metal components for machines were made manually, and therefore had an inherent inconsistency. But his machine manufactured stuff with consistent and precise measurements, opening up the world of interchangeable parts. Now components could be churned out in vast quantities with confidence that any of those components would fit where they needed to. The idea of mass production was foundational for the Industrial Revolution, and it led to the ability to make particularly thin and small things like screws, gears, springs, and wire.

Second came the discovery of how to make steel quickly and cheaply. While trying to improve the quality of the iron used to make guns, another Henry, Henry Bessemer (born in 1813), realized that iron could be heated to much higher temperatures by blowing hot air over it rather than using burning coal. This new process efficiently burned off impurities in the iron, after which the perfect amount of carbon could be added back in to create steel. Steel is stronger and harder than pure iron, it lasts much longer under wear, and it has a little flexibility. This makes it an ideal material for nails.

These advances enabled the fast-moving, hard-pressing machines of the nineteenth century to be made, culminating in our ability to manufacture large drums of steel wire, which was then used to make cheap nails.

These wire nails were thin and round, and at first they didn't gain favor with skilled joiners, as they had less holding power than square nails. Eventually, however, their much lower price won out, and production soared. In 1886, 10 percent of nails in the United States were wire-cut. By 1913, that figure was 90 percent.

Today, a nail-press machine can make over 800 metal nails a minute. There are several different types of nails available: smooth wire nails (the most common); nails with coatings such as zinc, which offer more corrosion protection; and nails with slight threads or texture, called barbed nails. But they all start life as steel wire rolled up on a drum, in just the same way as those created by the early twentieth-century machines. Typically, the wire diameter is 6mm, which is too large, so it's stretched by a series of spinning drums that pull on it to make it thinner. This thinner wire is then cut into rods. To convert each rod into a nail, two things happen. First, a blade presses one end into a sharp point, and then another machine applies a large amount of pressure to the other end to form the head. And that's all. There's no heating or bicep-building hammering involved. You could say that we've nailed it.

What do you see when you look at a nail? If you're an engineer, it's not a solid-looking, seemingly inert object, but the focus for a variety of forces that batter, push, pull, and shear it. These forces are what engineers through time have had to consider when using nails to fix things together and getting them to stay fixed.

To install a nail, you strike it with sharp blows to drive it into its receiving material. The sharp end ensures the nail can puncture the surface without doing too much damage. This is because the smaller the area the force acts on, the greater the pressure, so the point on one end of the nail channels the banging force from the hammer effectively. When you stand in grass wearing stilettos, you sink for the same reason.

As well as the force applied via the nail to the receiving material, there's the force received by the nail itself. The last time I put up a picture, the nail bent as I hammered it into the wall. You might think this is because I was hitting the nail slightly out of line, so the forces weren't going straight down its body. But that wasn't the only problem. In fact, I wasn't hitting the nail hard enough, which illustrates another characteristic of metal that makes it a good material for nails. It seems counterintuitive—surely the larger the force, the more likely the shaft is to buckle out of shape? This is indeed true of large buildings or bridges, where the weight of the structure bears down on the skeleton over a long period of time. A nail, however, behaves differently because of the way the force is applied to it and how it reacts to that force. Here, we're not talking about a long period of time: when you hammer a nail, the compression forces are huge, sending shockwaves through its body, but the duration of the loading is just a fraction of a second. Smash the nail hard enough and it doesn't have time to buckle. Part of the reason for this is the strange behavior of metals when loaded. The load at which they deform can depend on how quickly that load is applied: the quicker it's done, the more force a metal can resist without failing.

Once that nail is hammered in, it is friction that holds it in place. Friction is the force that arises when two surfaces are sliding, or trying to slide, against each other. If you try to pull apart two blocks of wood that have been nailed together, the wood fibers grip the shaft of the nail. The nail feels a force trying to rip it apart along its length, and we call that force tension. Your experiment can now fail in one of two ways—either the nail stretches and splits in half because the tension force is too large for the nail, or the nail comes loose because the friction force is overcome. The force it would take to stretch the nail is much larger than the friction forces on the surface, so we don't have to worry too much about the former. It's the friction with which we need to concern ourselves.

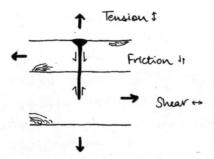

Forces experienced by a nail

The amount of friction between the nail and the block depends on the area of the two materials that are in contact, and their roughness. Wood can be an inconsistent material. Trees are living things that grow taller and thicker, have growth rings, and gain and lose leaves. Once a tree is chopped into wood, the hardness of its layers, moisture content, direction of the grain, temperature, and humidity of its surroundings all affect the material—and these things can change over time. All these factors influence the friction force as well. These are some of the issues that screws overcome, as we'll see.

The other force that nails experience is called shear. If you imagine the bottom block holding the nail stays stationary, but the top block moves sideways, then the nail can be deformed across its body; this is the effect of shear. The amount of shear a nail can resist depends mainly on its material and area of cross section—the stronger the material and the larger the area, the more shear it can resist.

The slight flexibility or ductility that steel displays as a material means that it is really good at absorbing these pulling and shearing forces. Other materials used before steel had limitations—wrought iron was too soft, and cast iron was too brittle—so that when they

were exposed to these forces, they were liable to excessive deformation or snapping. Nails made from these materials did the job
for centuries because they were so big, and thus strong enough to
prevent failure. But to make thin wire nails work, we needed to
wait for steel.

Small as it is, a nail driven into a wall to hang up a picture
nonetheless offers a good example of the forces that form part of an
engineer's considerations on a daily basis. The cables that hold up
the deck of a large bridge, like the Golden Gate Bridge, are being
tugged by the weight of the deck and the cars that drive over it, creating tension. And the beams that form the deck experience shear
because of their own weight and the load they support. The same
fundamental forces are at the heart of engineering, whatever the
scale of structure.

One large structure that shows the forces at work on nails, and
the solutions engineers came up with to deal with them, is the 600-
tonne warship *Mary Rose*—a favorite of the much-married English
king Henry VIII—which sank during the Battle of the Solent in
1545 and rested on the seabed for several centuries before being
rediscovered and put on display in Portsmouth, on the southern
coast of England. A large portion of the structure has been lost
to the waves, and the hull now gapes open like part of a decay-
ing ribcage. But it is still impressive to see how much remains,
especially given the chaos of her sudden plunge to the seafloor and
the centuries of buffeting by the shifting Solent tides. At her base
is the keel, the ship's backbone, which would have once sat at its
deepest point in the water. From the keel, ribs emanate upward,
forming the structural skeleton of the hull. Inside were four decks
made from beams and planks, the ends of which were held up by
L-shaped blocks known as "knees." (Maritime engineering has a

vivid vocabulary all of its own.) And protruding from the planks and knees are a series of wooden nails, or "treenails," the fasteners that held the ship together for centuries.

Treenails are cylindrical rods of wood, which are generally much longer and thicker than their metal equivalents. The treenails of the *Mary Rose* were up to half a meter long. Treenails don't have sharp points, a feature that we normally associate with nails, because wood simply isn't strong enough to pierce through another piece of wood. Instead, an auger (an early version of the modern electric drill) would be used to create a slightly undersized hole in the planks of wood first. The end of the treenail would be dipped in animal fat to help it slide into the hole, and finally driven in with a long-handled, heavy hammer. To lodge it in more firmly, the end was sometimes slightly split and filled with caulking, or fibers painted with tar, to create a subtle bell shape.

Wooden nails were well suited for ships because ship structures needed constant repair and replacement, and you could saw through them easily. Also, wood swells when wet, so the nails tightened up when the ship was at sea. But in some critical places, like the knees, the treenails weren't strong enough to resist the large forces, so iron rods were inserted in these junctions in addition to the treenails to provide extra strength.

For thirty-three years, the treenails held on tight as this immense ship sailed the seas, withstanding wind, waves, and battles with French fleets. Her end remains a mystery: as she approached enemy ships, she suddenly heeled over, took on water, and sank so rapidly that almost every crew member was lost. Eventually, the location of the wreck became uncertain, even though parts of it were occasionally spotted at low tide, but from the mid-1960s onward, a team of divers began an organized search of the Solent for the *Mary Rose*'s remains, finally locating them in 1971. By then, much of the iron had long decayed away in the salty water, so it was the wooden nails that still held the ship together. Any remaining iron was removed when she was retrieved to preserve the integrity of the wood. Now, as the body of the *Mary Rose* has dried and shrunk, the treenails—which were once flush with the surface of the wood—have emerged, poking their heads up, as if seeking recognition for their role in keeping her intact.

The *Mary Rose* was one of the first purpose-built warships. Two hundred and twenty years later, another warship, Britain's HMS *Victory*, was floated out into Chatham's Royal Dockyard. She was the largest timber warship of her time; at least 2,000 oak trees had been used in her construction, and thirty-seven individual sails were needed to guide her. Technology had improved by the time HMS *Victory* was constructed, so while large treenails still connected some of her parts, a vast array of metal fasteners also held her together. Here, too, we find an exotic, evocative vocabulary. The connectors in HMS *Victory* include the coax (a short, wide peg made of lignum vitae wood), the clench (immense copper rod that bound key structural elements), the deck dumps (thick, long wrought-iron rods that were curved, probably to join the deck structure to the hull), the lantern nail (a wrought-iron nail with an L-shaped head, so lanterns could be hung from it), the forelock bolt (whose tapered point had a hole in it, through which a wedge of iron passed to lock the bolt in place), and the Muntz metal spike (a large brass alloy nail with a sharp point, named after its creator, the Birmingham industrialist George Muntz).

By this time, engineers were more thoughtful about the materials they employed: copper connectors were used under the waterline, for example, because, unlike iron, copper doesn't react badly with seawater to damage wood. Muntz metal, which contained copper and iron, had the benefit of being cheaper than pure copper, with some anticorrosion properties. Iron was used above the waterline and inside the ship.

HMS *Victory* was state of the art when it was launched in 1765, and led fleets in the American War of Independence (1775–83) before achieving fame as Vice-Admiral Nelson's flagship in the Battle of Trafalgar in 1805. But this was a restlessly inventive period for engineering, and older methods of construction were quickly swept aside as the Industrial Revolution gathered pace. HMS *Victory* was one of the last great wooden warships to be made in the UK; the future would be vessels built largely of iron that required different sorts of fasteners. You can still see this monument to wooden shipbuilding in dry dock in Portsmouth, at the National Museum of the Royal Navy. If you visit, as well as marveling at the ship itself, be sure to check out the display case containing a motley collection of her fasteners—evidence of the often hidden but ingenious ways in which engineers join materials into complex, mobile structures.

The presence of wooden treenails in the *Mary Rose* hints at the fact that, for all their usefulness, metal nails aren't always the best choice for the job. Sometimes, this is down to simple necessity. Japan is not a country rich in iron ore, so the metal—which had to be laboriously extracted from iron sands—was mostly reserved for making their legendary swords. From the fifth century onward, Japanese temples and pagodas were constructed from pieces of wood carved to slot together in a precise sequence of complex, interlocking joints, with no metal connectors. The flexibility of these wood-only joints also made them better suited to surviving the earthquakes to which Japan is prone, allowing the energy to dissipate as the joints rattled.

In other situations, however, an engineer might be faced with circumstances in which the nail, for all its skill, just isn't up to the job. One fundamental drawback of the nail is its dependence on friction in order to grip securely. This means the nail has to be reasonably long, and its whole body needs to be encased by the materials being joined together, otherwise not enough friction will be generated to hold it in place. Vibrations and repetitive movement can overcome this friction, making nails come loose. In a piece of engineering likely to be subjected to continual shaking, nails may not be the best option.

Few things shake more than an aircraft. I'm a poor flier—being thousands of meters above the ground, with only very thin metal separating me from the air outside, makes me nervous. But I'd be more nervous still if I didn't know that engineers had thought up a different kind of fastener that can handle these vibrations.

Given my fear of flying and the nail's unsuitability for aviation construction, I was a little disturbed to discover that parts of the early planes were actually held together with nails. I was also full of admiration for the extraordinary crews who flew in them, such as the 46th Taman Guards Night Bomber Regiment. One of their navigators, Polina Vladimirovna Gelman, began learning to fly gliders during the 1930s, when she was a teenager, but her short stature (or, you could argue, a large aircraft not designed inclusively) meant that when conducting a maneuver her instructor had shown her during her first glider flight, she slid down in her seat in order to reach the rudder pedals and disappeared out of sight. Gelman was told not to come back. But when Germany invaded Russia during the Second World War, she heard that an aviator named Marina Raskova was forming an all-women flight regiment. The aircraft was still not compatible with Gelman's frame, so although she yearned to become a pilot, she ended up training to become a navigator instead.

Gelman and her pilot flew a plane called the Polikarpov Po-2. Under the cover of night, they and the rest of the squadron targeted German trenches, supply lines, and railroads, switching off their engines as they approached, coming in on a silent glide before dropping their bombs. In the silence of those glides, you could hear the wind whistling through the structure of their wings, and soldiers below likened this sound to witches on brooms, derisively calling them *Nachthexen*, or Night Witches.

Initially underestimated by their male counterparts, the women pilots more than proved their value as they flew, often sleep-deprived, in harsh weather conditions. Considering that most Allied pilots only carried out between thirty and fifty missions before returning home, it is incredibly impressive that Gelman completed 860 missions as senior lieutenant of her regiment. She is the only female Jewish Hero of the Soviet Union, the highest distinction bestowed on any Soviet citizen or foreigner (being Jewish meant that she wasn't considered Russian), and she later received the Gold Star medal for her service.

I went to see a Po-2 at the Shuttleworth Collection of airplanes in Biggleswade, England. As aircraft go, it's cheap and crude, but rugged and reliable (this particular model is still in good enough shape to fly). Just 8 meters long, it's a biplane, meaning it has two sets of wings, one above the other. But the most noteworthy feature, for me, is the fact that the Po-2 was made from wood, which meant nails had a role to play in its structure and construction. The main body—the fuselage—was built of four long, stiff beams of wood—top and bottom, and one on each side—with vertical wooden beams holding them together to create a long rectangular frame. Between each vertical beam was a diagonal piece that braced the tube, making it stiff. This primary structure was connected with strong steel plates and bolts, or glued together with nails as reinforcement. The outermost painted surface of the fuselage was made of linen glued to a secondary network of thin, curved wooden ribs. Nails were used to clamp linen to wood while the glue dried, and then often removed to reduce weight.

Wood had been used for the construction of planes for the First

World War, but aviation engineering had evolved since then. So, while the fuselage of early Second World War planes, like the famous fighter Hawker Hurricane, still had the same structural form as in the previous war, they now generally had only one set of wings, and were made from steel instead of wood. This was a transition period, when aircraft often had a steel primary structure, but used wood and fabric to form their outer shells. In these planes, nails still made an appearance in clamping down the fabric, but the steel plates and beams were made from thin sheets that offered little material for the body of the nail to be hammered into. There was no way a nail could hold such thin things together. A different fastener was needed. Fortunately, the engineers of the ancient world had already figured out a fastener that could do the job—the rivet.

Like wire nails, rivets have a shaft that is cylindrical but is thicker than that of a nail. Unlike wire nails, rivets do not have a sharp end. Instead, they have two domed heads, making them look like tiny dumbbells. Look around you, and you might see Industrial Revolution–era railway bridges and buildings dotted with distinctive rows of domed heads tying structural beams and columns together. Today, rivets are out of fashion in such structures, rejected in favor of bolts, which we'll come to. But engineers in the aerospace industry still extensively use rivets, which win out over space-hungry and heavy bolts any day. Shipbuilders, who also now work in metal rather than wood, favor rivets. But these industries are simply beneficiaries of the rivet. They didn't cause engineers to invent it. For that, we have to go much further back in time.

Two types of Roman mail armor—*lorica hamata* and *lorica segmentata* (which comprised metal hoops fastened to leather straps)—used iron rivets. Rich, the blacksmith who taught me to make nails at Much Hadham forge, explained how this had come about. Inspired by woodworking techniques, smiths tried to use mortise and tenon joints to connect large pieces of iron together. These joints, which comprise a tenon, or protrusion, at the end of one piece that slots into a recess, the mortise, in another piece, work well with wood, which can be carefully carved, but the nature of iron and the way it is shaped is such that you can't get smooth surfaces like you can with wood, meaning you can't engineer enough contact between the mortise and tenon to create the required friction. Also, friction alone wouldn't be sufficient to hold large, heavy pieces of iron together. So, they created longer tenons that poked through the mortise, heated up the end of the tenon, and then hammered them back to create a head. This is the basis of the design of a rivet.

But the rivet actually predates this. There is speculation that the Egyptians were using rivets as far back as 3000 BCE, not long after they were making nails. The Museum of Fine Arts in Boston holds evidence, in the form of a bronze jug, dating back to 1479–1352 BCE, which was found in an ancient tomb in Abydos, Egypt. It's a beautifully worked piece, with a squat cylindrical neck above a bulbous body, its surface a mottled mixture of deep red and brown hues, with little dimples showing where the bronze was hammered into shape. On one side is a handle in the form of a lotus flower, which is attached to the main body with three rivets.

For engineers ancient and modern, the virtue of the rivet is that,

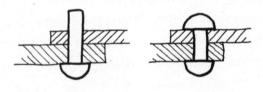

Rivet: before, and after, installation

unlike the nail, it doesn't rely on friction to stay in place; instead, it holds things together because of its shape. If you try and pull apart the two sheets of metal that a rivet is holding together, the inside faces of both domes resist the force as the shaft goes into tension. To fight forces in this way, the rivet has to change during installation (again, unlike the nail, which remains the same). There are two types of rivet—hot and cold—both of which begin life as a cylindrical shaft with a hemispherical cap at one end. Prior to installation, a hole is drilled or punched through the two layers that need joining.

To join large pieces of structure with big forces, like the nineteenth-century bridges I admire, hot rivets provide the strength. The rivet stoker would have heated it up in the forge, and then tossed it quickly to a rivet catcher, who would grab it with tongs and pass it through the predrilled holes (I'm glad that chucking glowing-hot pieces of metal across a construction site has stayed in the past). Then another riveter would begin beating the protruding shaft with a hammer containing a "rivet snap," a heavy piece of metal that had a dome carved out of it. Sharp blows would smash the shaft into another dome shape, so that it now had two hemispherical heads holding the materials in place. As the rivet cooled, it would shrink a little, meaning that it exerted an even stronger clamping force. Hot rivets were made from hard materials like iron or steel, so they needed to be heated up in order to form the second head. Nowadays, the process to install all these sorts of rivets is mechanized.

Alternatively, an engineer could choose to use cold rivets. They don't need heating because they are made from softer metals, like aluminum, which means they can be more easily shaped. They are often hollow, and far smaller than their hot counterparts, making them far lighter as well (if not as strong). These sorts are used in aircraft, where minimizing weight is key to reduce the amount of fuel needed.

Planes like the Hawker Hurricane had hundreds of hollow rivets holding them together, alongside screws and bolts for the heavily loaded junctions of the frame. But rivets also enabled the fundamental design of the airplane fuselage to change. Hurricanes had two sets of structure made from steel and wood forming the fuselage—an inner primary structure and an outer curved frame, to which the skin was attached. But the Spitfire—the iconic Battle of Britain fighter plane, with its distinctive elliptical wings and streamlined body, which could reach speeds close to 600 kph—was radically different.

When I peered inside a Spitfire, standing not far from the Po-2 at the Shuttleworth Collection, the first thing I noticed is that the inside was almost empty. Just inside its outer skin, I could see a series of curved aluminum arches one behind the other, forming the main structure of the fuselage. Along the edge of each beam was a narrow rim, which lined up with the outside skin. It was these two layers—the rim of the beam and the outer skin of the fuselage—that were clamped by thousands of rivets. Both layers of aluminum were extremely thin. They needed to be held together in a way that meant the two layers acted as one, for if they slipped around, they wouldn't act as a single, strong entity. Rivets, effective at resisting the shear forces across the shaft, were perfect. The key point here is that whereas the Hurricane needed two frames (structure and skin), the Spitfire had one. This new form of fuselage created a light but robust high-performing aircraft that was ahead of its time.

Although rivets usually have the two domed heads I described earlier, the designer of the Spitfire, R. J. Mitchell, wanted the head on the outside of the aircraft to be flat to allow a smooth airflow over the plane, so it could fly faster. But flush rivets, as flat-headed rivets are known, are more expensive to manufacture and take longer to install, so it was decided to verify whether flush riveting genuinely made a difference to the Spitfire's speed. The methods they employed to test this were unusual. Engineers glued a split pea to the head of every flush rivet on the plane (making it look, according to one source, like it had a "chickenpox infection"), then flew it and noted the speed. More test flights followed, in which the split peas were removed in stages and the results noted. This ultimately vindicated Mitchell's choice of flat heads: data showed that domed rivets would reduce the top speed of the fighter plane by up to 35 kph.

In the postwar decades, passenger planes were generally made from aluminum with a structure very similar to the Spitfire. The fuselage and wings experience large forces that flex and twist them, and they also are hit by sudden gusts, large temperature variations, and strong vibrations. Rivets are an absolutely vital component of an aircraft; a single Boeing 737 can have over 600,000 rivets and bolts giving it integrity.

The advent of cheap and quick extraction and manufacturing methods for sheet aluminum in the early twentieth century opened up a new world of design in the aerospace industry, but without rivets we couldn't have envisioned the large planes we travel in today. We would still be dependent on the intrusive frames formed inside the outer shell of the Po-2 and the Hurricane, which restricted their passenger capacity. The open, riveted tube structure of the Spitfire freed up space inside planes, allowing us to create large and light modern aircraft that carry hundreds of people and tonnes of heavy goods.

Complicated machines like the Spitfire could be made thanks to the era of mass manufacture and assembly lines that began at the end of the eighteenth century and expanded in the nineteenth.

Engineers such as brothers Job and William Wyatt, and others like Henry Maudslay (the inventor of the metal-cutting lathe) and Joseph Whitworth, revolutionized manufacturing methods so that stuff could be made to tiny and precise measurements, in large quantities, and to specific standards. That stuff included the screw.

The screw, like the rivet and the nail, is an engineering solution to fastening things together. It is similar to a nail in that it has a long shaft and a head, but instead of being smooth, its shaft has a helical thread wrapped around it. It's also similar in that both have a pointy end that punctures the material you want to join to something else. But unlike a nail, you rotate the screw so that the thread cuts into the material and burrows the fastener. And, again unlike a nail, the screw physically clamps stuff together between its threads with a mechanical hold.

Friction is still important in screws. It acts between the threads and the material it's been installed in, and stops the screw from rotating and coming loose. If you try to pull apart the two pieces being held together, rather than a straightforward tension in its shaft, there is also the force of the material pushing against the thread all along it. Screws also experience a completely different force to the nail during installation; rather than a swift bang, it feels a steady twisting or torsion force as it turns under a screwdriver. The material from which the screw is made needs to be strong enough that it doesn't deform when twisted.

You might consider the screw the better fastener—a sort of grown-up nail. It's true that screws are generally made from harder materials than nails, because they need to be strong enough to be turned in without deforming. They tie things together tightly thanks to their threads, and they resist tension (pulling forces), so you can hang things from screws without worrying too much about them coming loose. The helical threads can be made really compact, and their shafts very short, meaning that screws can hold tiny things together, like parts of a wristwatch, or thin sheets of metal. And it's much easier to remove a screw after it's been installed; you simply rotate a screwdriver the other way.

At the same time, because they are harder, screws are more prone to breaking in shear, whereas nails are more ductile or flexible, so they don't snap as easily. Until the design of screws incorporated a shape in its head (like a + or − sign into which a screwdriver slots), they were tricky to install. Before Maudslay's machine, they were expensive and laborious to make—the thread for every individual screw was painstakingly hand-cut. This meant screws simply weren't widely available, if at all, in Europe before the fifteenth century, and only became a cheap and widely used product during the Industrial Revolution.

Nevertheless, the concept of the screw—or, more specifically, of a helical thread wrapped around a shaft—has long been a feature of engineering, one that goes far beyond the act of joining two things together. Ancient Egypt had an irrigation tool that is now called the Archimedes screw. Imagine a hollow wooden tube, inside of which is a long, threaded cylinder. Set at an incline, one end is dipped into a river or lake, and the other is higher up on land. As you twist the screw with a handle, the threads trap water inside the tube and the water is pulled up. This form of irrigation is still used along the Nile in Egypt and is arguably where Archimedes got his inspiration. Now, however, there are theories that this system was actually invented a few hundred years earlier, in the sixth century BCE, by engineers serving King Nebuchadnezzar II, to irrigate the famous greenery of the Hanging Gardens of Babylon. As far as we know, the water screw was the first appearance in history of a helix, a mathematically complex shape that was also tricky to make.

By the sixteenth century, the screw was included in a list of simple machines (along with the lever, the wheel and axle, the pulley, the inclined plane, and the wedge). The idea of "simple machines" was a concept first developed by the Greeks and then refined by Renaissance scientists and engineers, identifying tools with few or no moving parts that help us to work against forces, often by changing the direction of those forces.

For example, it's much easier to raise a heavy weight by pushing it up a slope (a technique used in constructing the pyramids) than by lifting it vertically. Or consider the ability of a child to lift her father using a seesaw, which is a type of lever. If he sits close to the support on one side, he can be raised up by pushing down on the end of the other side: clearly not something she could achieve using her muscles or body weight alone. In the case of the screw, its helical threads convert rotational movement to linear movement: so, in an Archimedes screw, the rotation of the shaft pulls water up along a straight line.

Skyscrapers couldn't exist in some parts of the world without this rotational-to-linear movement. The earliest towers appeared in places like Manhattan, where the ground was hard, and a thick slab of concrete was enough to spread the immense weight of the tower into the rock below. But cities like London are based on a softer and more variable material, clay, which is a fine soil made from particles that are transported and ground down by water. When it's wet, clay expands; when it's dry, it shrinks and can crack. It changes from season to season, and year to year, which is problematic for the foundations of a tower.

To solve this, engineers use piles, which are stilt-like foundations put into the ground to support structures. A skyscraper might be held aloft on immense shafts of concrete over 1 meter in diameter that plunge about 40 meters into the ground. They might seem a world away from the nail—and they are in scale—but they experience forces in a similar way. Like nails, friction piles work because of the forces acting at their surface, i.e., between the outermost layer and the soil.

However, constructing a pile presents a challenge to the engineer, because you need to dig a relatively narrow but very deep hole, into which wet concrete is poured and allowed to harden. Here's where the magic of a screw can be used, in a piece of machinery called a piling rig. Some of these machines have an enormously long vertical screw driven by an engine that rotates it into the ground. The soil gets trapped within its threads, so when the screw is withdrawn, it comes out with mud stuck to it, leaving behind a clear hole—just like a corkscrew extracts the cork from a bottle of wine. This hole is then filled with liquid concrete, into which a steel wire cage is lowered. As the concrete hardens, it binds with the steel, forming a strong and durable structure that effectively channels loads into the ground.

The era of precise and mass manufacture saw the birth of another fastener, one of which I'm particularly fond. It is now an expression meaning the fundamental little parts or elements that make something work, and it's what this book is named after.

Bolts are, in some ways, an amalgamation of screws and rivets. They have a long cylindrical shaft topped by a hexagonal head, so shaped to be tightened using a wrench. All, or sometimes just the end, of the shaft, is threaded—so far, it mostly resembles a chunky screw. To put one in, you need preformed holes in the steel beams or columns being connected—just as with rivets—and then

A nut and bolt

a single person can simply turn a nut onto its end. The nut is a hexagonal doughnut, with the inner surface of the hole threaded opposite to the shank of the bolt. Thanks to machines like Maudslay's lathe, the threads match up perfectly, allowing the nut to rotate onto the shank.

The bolt may look like a screw, but in terms of the forces and the way it works, it's more like a rivet. When the nut and bolt are tightened together, they clamp the pieces of metal they need to hold together. In large structures like bridges and skyscrapers, where relatively thick sheets of iron or steel need joining, screws wouldn't work because it would be almost impossible to physically cut threads into the metal sheets by turning a screwdriver. Hot rivets work up to a point. But they are made from a relatively soft form of iron so that the second dome can be hammered on site, a dangerous and laborious task. The process of tightening up a nut is rather safer than flinging hot bits of metal around. Bolts can be made from harder steel than rivets, making them much stronger. A single bolt of the type commonly used in construction today, which is just 20 mm in diameter, can take pulling loads of around 11 tonnes. This is about the weight of a double-decker London bus.

But this is not the whole story. When I reconnected with engineer and self-confessed "bolt nut" Omar Sharif, who checked the forces in thousands of bolts in The Shard, he said this book should be called *Nuts, Bolts, and Washers*. He has a point. Washers, which are thin, flat rings of steel, are essential to the working of bolts. They sit between the nut and the steel member being connected, and are responsible for spreading out the clamping force of the nut. Without a washer, tightening up a nut can create tiny cracks in the beam or column and weaken it. In Omar's words, the washer is as important to the bolt as Alvin is to the Chipmunks—one just doesn't work without the other. (After much consideration, I decided not to rename the book, after all. *Nuts, Bolts, and Washers* doesn't have quite the same ring to it.)

The bolts used to connect the top portion of The Shard, the Spire, presented a very particular engineering problem. The gusts of wind that blow on the Spire are channeled down through its steel frame into the concrete backbone of the main tower. The frame is like a skeleton, which keeps it standing strong, and the most critical parts are its joints—the connection points between the bones. Complex configurations of bolts and welds bind the beams and columns, holding the building aloft. Since the Spire was exposed to the elements and its structure visible to visitors in the viewing gallery, the bolts had to be strong enough to resist the wind forces, durable enough to resist wear from the weather, easy enough to install high up in the air, but also beautiful.

Given the amount of effort that went into this, you won't be surprised to know that when I visit, I resist gazing at the impressive views of London spread out before me. Instead, I look up lovingly at the bolts—which were inspired by nails—that hold the Spire together. (Omar assures me he does the same.) These bolts are strong, beautiful, carefully designed pieces of engineering with a long history, each of which would fit safely inside the palm of my hand.

Wheel

An alarm clock rings. I roll over in bed, switch it off, and check the time on the dial. Trudging to the bathroom, I turn on the tap, then flick my electric toothbrush into motion. Once washed and dressed, I move to the kitchen, swing open the fridge door looking for milk, which I pour into a pan, stirring in oats. Maybe today's the day I treat myself to a homemade smoothie in the blender, too. Breakfast over, I turn the doorknob, leave the apartment, board a train.

A typical morning for a lot of us: mundane, even. But next time you do it, notice just how much of the beginning of each day is filled with rotating things. And so it continues: from the vehicles that move us, to the sphere in the tips of our ballpoint pens; from the pulley that lifts loads in a crane, to the gyroscopes that stabilize the satellites that tell us our location and the time. Ask people what is the best/most influential/most enduring (take your pick) invention of all time, and I'd guess that most will at least consider the wheel—even if they try to come up with something more original, because the wheel is such a standard response.

Even though the wheel is so familiar to us—we feel like we know what it is and what it does—it still holds some surprises. For a start, the reason the wheel is generally considered the best invention ever is because of its world-changing impact on our mobility, but in fact the wheel was not invented to move us; it had, at first, a very different purpose. What's more, we think of the wheel as something that has been unchanged over millennia—a round thing with an axle around which it can rotate—to the point where it's become a cliché: *don't reinvent the wheel.* (Ironically, in 2001, an attorney in Australia managed to obtain an innovation patent for a "circular transportation facilitation device." He was trying to highlight deficiencies in a new patenting system by reinventing the wheel.) But I don't agree with this cliché. Over the last 5,000 years, the world has changed drastically, and in that time, we *have* reinvented the wheel, over and over again. Not only have we played with its form and used new materials to make it, but we've also completely changed *how* it's used and *what* it can do for us—that's reinvention in my book.

For a number of inventions—winged flight, Velcro, sonar—our inspiration has come from nature. The wheel, however, is a human triumph. Armadillos curl up into a ball and roll, as does tumbleweed; and dung beetles form poo into spheres that they can then push around easily—but the idea of a spinning object interacting with a nonspinning object to create a device has no natural precedent. It was truly . . . revolutionary.

Less of a human triumph was the shapeless splurge of clay I created one day in a studio, which then continued to spin mockingly in front of me, as though it wanted to display the ugliness of my efforts from all sides. I had thrown the clay onto a rotating wheel and attempted to shape it with my hands. It took just twelve seconds for me to realize that my skills as a potter were going to need some work. Other humans, though, have been artfully fashioning objects with clay since at least 29,000 BCE. While clay might not be the best foundation material for tall buildings in the modern world, its malleability made it the perfect material for our ancestors to work with. Initially, they molded clay vessels from scratch by hand. Then they developed the coiling method: laying down long rolls of clay in a stacked spiral and using fingers to smooth down the walls. This, though, was a slow process, and as humans began to establish more permanent settlements, and to grow, store, and cook food, they needed a quicker way to make large numbers of sizable, good-quality pots.

It is for pottery that the wheel was originally invented. The oldest wheels we've found appeared around 3900 BCE in Mesopotamia. The potter's wheel was a heavy, large disc made from baked clay or wood. Its upper surface was flat, but the underside had a bulge, which was nestled into a fixed piece of wood or stone whose top was curved. Potters would set the upper disc spinning with their hands, and its weight meant that it would keep turning for some time.

For a while, pots were still made by coiling, but they would be placed on the wheel to smoothen and perfect quickly, producing far more consistent pieces. Later on, mechanisms to drive the wheel with foot pedals freed potters' hands, allowing them to focus on using them to form the clay into shapes, and the consistent spinning motion eventually led to the technique of throwing a lump of clay onto the center of the rotating disc, just like I did. Unlike me, however, skilled crafters could now make pots with consistent sides very rapidly, creating enough vessels to satisfy settlements' increasing demand for them.

Potters in Mesopotamia had harnessed rotational movement about a fixed point for the first time. Although we often think of the wheel as the invention that predates all others—for which I blame Fred Flintstone and his caveman car—we were making jewelry,

wine, boats, and musical instruments (which are all pretty impressive feats of engineering) long before we thought up the wheel.

The idea of using circular motion to take you forward linearly
was a leap of imagination. Someone really did have to reinvent
the wheel—or, at least, the way in which it was used. And it does
seem to have been a leap: most inventions evolve over time. A naturally sharp piece of rock, for example, inspired us to sharpen other
rocks into tools, and gradually we began to attach them to handles,
long poles, or arrow shafts—but with the wheel and axle, there's
no such evolution in getting to its basic form. It either works or it
doesn't. The people who invented it must have had advanced carpentry skills in order to carve the hole at the center, thick-trunked
trees from which the disc was carved, and the need to transport
heavy things across relatively flat land (few inventions materialize
without having the need for them first). And apart from this perfect storm of dexterity, topography, and technology, they needed
some really innovative thinking, because the axle—which makes
the wheel a practical object—is a complicated piece of engineering
in its own right.

At its most basic, an axle is just a rod that passes through a
wheel. They can come together in two ways: either the axle is fixed
to the vehicle and the wheel spins around it, or the wheel and axle
are fixed together, and they both rotate. The axle needs to be substantial enough to bear the load of the vehicle. It also needs to fit
loosely enough so that the system can rotate, but snugly enough
that it won't rattle. If the axle is too thick, the amount of friction
increases, slowing down the system and shortening its life through
wear. The surfaces of the wheel and the axle that are in contact also
have to be smooth and near-perfectly curved, and people struggled
with the careful carpentry needed until metal tools like chisels

became common. For all these reasons, in my expert engineer's opinion, the axle in Fred Flintstone's car would never have worked.

Because of the complexity involved in the invention of this system, some historians argue that it is unlikely that it was independently invented multiple times. They suggest that, once invented, the wheel and axle found their way across the Eurasian continent very quickly. After all, the advantages were clear: for centuries, people had depended on animals and/or sledges to move things around. But animals tire and need food and looking after, and a sledge is effective where the ground is flat and icy, so it can glide, but otherwise there is a lot of friction to overcome between the legs of the vehicle and the ground, making it cumbersome. Round it might have been, but the wheel definitely had the edge.

The quick spread of the wheel makes it challenging to pinpoint its origin. Archaeologists have found clay tablets featuring wagons from Uruk, Mesopotamia, from the mid-fourth millennium BCE, which is about the same time that a ceramic pot found in Bronocice in modern-day Poland appears to depict a wheeled vehicle. In areas around the Danube and North Caucasus, clay models of wagons have been unearthed. Wheeled toys also made an appearance in the Americas precolonization, suggesting that the wheel might have been invented independently there, but there's no evidence of full-sized wheeled vehicles, maybe because the key civilizations were situated in a lake surrounded by mountains (the home of the Aztecs) and steep ranges (the Incas), which meant that animals made better modes of transport.

For physical archaeological evidence of an actual wheeled vehicle, we have to travel east of the city of Stavropol, in the North Caucasus region of Russia. Here, archaeologists found a site containing tens of thousands of ceremonial burial mounds. They had been dug by the local people in the fifth millennium BCE and then reused in the fourth millennium BCE by the Yamnaya community, who also added more graves. One of these contained something intriguing. At the bottom of a narrow, deep, catacomb-type shaft, archaeologists unearthed the skeleton of a man buried in a seated position on a four-wheeled wagon. Although the wagon had severely deteriorated, it has been dated back to between 3356 and 3033 BCE, making it one of the earliest in existence.

The wheels were solid, and each was made up of three boards of oak, connected with wooden dowels or pegs (similar to treenails— another example of how foundational nails are to engineering). The reason that simply cutting a round slice of tree trunk wouldn't work well is because the tree's natural grains make it stronger in some directions and more prone to splitting in others. Repeated use would affect the weaker direction more, causing the wheel to deform. The Yamnaya are also believed to be one of the earliest peoples to domesticate animals, which in turn would have pulled the wagons—so animal husbandry has also influenced this invention, because without animals the carts wouldn't have been as useful.

The wheel and axle, and the wagons it led to, changed food production. When our ancestors started agriculture, large teams of people were needed to trudge up and down the fields, tilling the land to grow food. To travel, they relied on animals or their own two feet. But with the help of an ox or horse and a wagon, a single family could reap plentiful crops from the same land and transport them across large distances before they spoiled. The wheel liberated in other ways, too. Previously, the Yamnaya had lived in small villages clustered around water. Now they became explorers. Riding these unfamiliar vehicles gave the Yamnaya military advantages over the existing settlers they encountered, and they expanded their home into the vast steppes and beyond.

As they rolled outward, the Yamnaya also rolled out their culture. Nearly half the world speaks languages that have descended from the language believed to be spoken by the Yamnaya, which is called Proto-Indo-European, and birthed languages as diverse as Sanskrit, Greek, Latin, Pashto, Bulgarian, English, and German. They shared their skills in animal domestication and metallurgy with those they encountered. They may even have unwittingly brought the Black Death to Europe: geneticists have found the responsible bacteria in the remains of human teeth in the region from which they originated. I can't help but wonder what Europe and Asia might look like today if the Yamnaya hadn't created their wagons.

Turning the potter's wheel on its side to convert the wheel from a vessel-maker into a destination-taker was only the first reinvention of this rotating invention. The early wheels that expanded the horizons and changed the lifestyle of the Eurasian people were solid, but a new innovation would transform them into something lighter and faster.

I speak Hindi, a language derived from Sanskrit, which can trace its origins back to the Yamnaya, so the journey of the wheel is intertwined with the history of my mother tongue. When I was growing up in India, the motif of the wheel was a constant presence. Early versions of the nation's flag designed during the fight for independence from British colonizers had at the center a spinning wheel, or charkha. One of the ambitions held by leaders of this Swadeshi Movement, which included Mahatma Gandhi, Aurobindo Ghosh, Rabindranath Tagore, Lala Lajpat Rai, and so many others, was to break free from economic dependence on the British and empower the colonized people of India. Gandhi was well known for making his own clothes and encouraged this action of peaceful civil disobedience, because making cloth locally was against the rules of the British Raj.

The spinning wheel, which traveled with him and was featured in iconic pictures of the freedom struggle, became a symbol of Indian independence. Today's flag has the chakra, a wheel in navy blue on a central white stripe, sandwiched between stripes of saffron and green. This wheel, which featured prominently in ancient Indian iconography, art, and architecture, dates back to the era of Ashoka, an emperor of the Maurya Dynasty who ruled much of the Indian subcontinent between 268 and 232 BCE and spread Buddhism across Asia.

Unlike the wheels on the wagon discovered in the Eurasian steppes, the charkha and chakra both have spokes. It was an insightful piece of design that transformed the wheel. The Sintashta, from the northern Eurasian steppes, and the Indo-Iranians may have had wheels like this at the end of the third millennium BCE. Spoked wheels from the second millennium BCE were discovered in Tutankhamen's tomb in Egypt, as well as in Mitanni records (modern-day Syria and Turkey). These wheels were constructed from wood, with a hub at the center, and a varying number of rods radiating out toward a circular rim. (The Ashoka chakra has twenty-four spokes, each of which symbolizes a principle for leading a good life, echoing the number of hours in a day.)

Hindu deities are often depicted riding their *vahana*, or vehicle, and the sun god Surya's is usually a chariot with spoked wheels that is drawn by seven horses. It's not surprising that even the divine would have preferred the spoked version; after all, they are much lighter than their solid counterparts. Although more complex to make, spoked wheels enabled much faster vehicles than the trundling solid-wheeled wagons. The Romans raced with spoked-wheeled chariots, and the Greeks used them in war.

While they were undoubtedly an improvement, spokes had drawbacks at first, which needed to be solved by further engineering innovation. Since spoked wheels were constructed from many pieces of wood, they were prone to falling apart after being jostled around for a while. However, in Europe, as metalworking skills began to spread in the Iron Age (1200–600 BCE), a flat metal hoop was added to the outside of the wheel's rim to give it strength. The clever thing was that the iron "tire" was measured so that it matched the circumference of the wheel while the metal was hot, which is when it was wrapped around the rim. Then, as the metal cooled, it shrank. This had the effect of squashing the rim inward, pushing the wooden spokes firmly into the hub, which made the wheels far more robust and long-wearing.

So, we had figured out how to make wheels stronger and speedier. What would be the next development? It would take a thousand years for another radical rethink to take place.

In the early 1800s, aeronautical engineer George Cayley was working on constructing flying machines. He needed strong wheels that would absorb the big bouncing forces on landing. But they needed to be light, otherwise he would struggle to get the aircraft airborne in the first place. Originally, he used wood, with rims and spokes strong enough to take the compressive forces of heavy vehicles. But then Cayley started experimenting with metal, and reversing the way forces were channeled through. Instead of the spokes being squashed, which requires a strong material that won't bend and break, he stretched thin metal wire between hub and rim, so the system was held together because these wires were being pulled, or in tension. The resulting design, which he invented in 1808, was far lighter than its predecessors, a real boost for the lightweight aircraft he, along with many others, was trying to create.

Another difference between the wooden wheel and the wire wheel was its profile, which hints at how the materials respond differently to the action of forces upon them. To picture this, imagine cutting out a circle from a stiff card and holding it up vertically.

If you poke at its center, it is pretty good at holding its shape, but with a bit of extra force, it can warp. And if you do the same thing with a circle made from a sheet of paper, it deforms even more easily. However, if you cut out a wedge from your paper circle, then bring the edges together and tape them up to form a shallow cone, the structure becomes much more stable and less prone to deformation. While many wooden wheels were made flat—like our card circle—coachmakers in Britain often created their wheels with a slight "dish" in one direction to give them more stability over uneven roads and while being pulled by horses whose backsides sway sideways. Dishing the wheels in one direction worked well as long as the wheels were made from materials that were strong when compressed, like wood, and there were two wheels to an axle, meaning they could counteract each other.

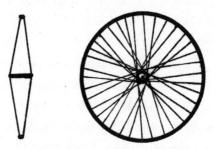

A wire wheel with its double dish, in profile and face-on

Ingenious as this is, it won't work for a wire wheel. Say you made a model of a wire wheel by tying together six pieces of string at one of their ends to form a hub, and then stretched and tied the other ends to a stiff ring (like a circular cookie cutter). If you push at the central hub, the strings would stretch and the hub would move sideways, which is not what you want in a wheel. And if you have a single dish, it's harder to deform the wheel when you poke at the inside face of the dish, because the strings are strong in tension, but push from the outside, and the hub, once again, just moves. So, to make the wire wheel work, a double dish was designed. If you look carefully at modern bicycle wheels, you'll see that at the hub, there are two sets of wire emerging that come together at the rim.

These double-dished, tension-wire wheels marked a very important stage in the evolution of the wheel—they were strong, flexible (so could be jostled without being easily damaged), and light. But even then, the wheels were arranged side by side, in pairs. Surprisingly, it took nearly a decade after the invention of the wire wheel for someone to think up perhaps the most significant innovation since the spoke—putting one wheel in front of the other.

Riding the Laufmaschine must have been a jarring and draining experience. Invented in Germany by Karl Freiherr von Drais in 1817, it had wooden wheels and a wooden frame, but no pedals, so you had to use your legs and run, just like Fred Flintstone did in his car. Nevertheless, the Laufmaschine was a big leap forward in transportation, being the first vehicle in which people didn't have to rely on animals to travel significant distances. Like a lot of inventors, Drais was ahead of his time: his design was derisively compared to children's hobby horses by the press, and poor roads made riding it unpleasant. Pedestrians were frustrated by riders who mounted the pavements, and they were seen as a nuisance, causing busy cities like Milan, London, New York, and Calcutta to ban their use.

Over time, people's attitudes changed. Denis Johnson, a British coachmaker, saw Drais's design and swiftly patented it in England before anyone else could. He called his machine the velocipede, and made some improvements, including providing larger and more stable wooden wheels lined with iron. He also introduced a drop-frame model for women. In the next few decades, pedals and brakes were added, and then, finally, wire wheels became the norm. In 1888, veterinary surgeon John Boyd Dunlop wrapped a soft tube around his son's tricycle wheels to give him a smoother ride, which led to the air-filled tire, making bicycles safer, easier to maneuver, and more comfortable.

Bicycles brought a huge change to the everyday lives of people: for the vast majority of the population, who couldn't afford horse-drawn carriages or early motor cars, this was the first time they'd had their own personal means of longer-distance transportation. Nurses and clergymen began traveling to rural areas and serving larger populations, and the post office started making daily deliveries to every household by the end of the century. Women were largely criticized and ridiculed for sitting astride the machines: Lillian Campbell Davidson, who wrote the *Handbook for Lady Cyclists* in 1896, recalled that "it was openly said that a woman who mounted a bicycle hopelessly unsexed herself." Despite such prejudice, the bicycle was a liberating piece of technology. N. G. Bacon, a leading advocate for women's cycling in this period, reflects that "the cycle allows us, in a perfectly womanly . . . manner to let ourselves go." And this seems to be true in more ways than Bacon might have imagined. The biologist Steve Jones ranks the invention of the bicycle as the most important event in recent human evolution, because owning a bicycle dramatically increased the geographical area that could be covered, and consequently the number of potential marriage partners, leading to a widening of the gene pool.

Fortunately, it's now generally acceptable for women to ride bicycles, but societal norms mean that becoming an engineer can still be a challenge for us. I find that much of the fashion aimed at appealing to engineers is designed with suit-wearing men in mind (pocket handkerchiefs and ties—so many ties), so I was excited to find some extremely nerdy earrings one day. Now my favorite pair, they are made of very thin, layered plywood, laser-cut into four spoked wheels of different sizes, the largest being jade green, overlaid with black and white ones. But there's something different about these wheels—rather than the smooth rims we've seen so far, these have serrated edges.

These serrations mesh perfectly with each other. If I manually turn one of my earring-wheels, the others rotate, too. I love them, not only because they are a visual reminder to people of my passion for engineering but also because they beautifully demonstrate how such an arrangement of wheels—or rather, an incarnation of the wheel called the gear—can pass on their movements and forces between them. These little wonders are hidden inside most machines, from clocks to cars to can openers to cranes; it's difficult to think of modern machines that don't have them.

Gears are wheels with teeth. Sometimes they're called cogwheels or even cogs, which is—confusingly—the name for the teeth themselves. The reason gears are so useful is because when you put two (or even more) side by side so their teeth overlap, there are three things you can do: change the direction of rotation, change the speed of rotation, and change the force acting at the rim of the gear.

Can openers are interesting little tools. They need to have two separate arms that come together to clamp the can, then later unclamp to release it. The blade needs to be turned to go around the circumference. Cogs help achieve these aims: two gears (one on each arm of the tool) come together with their teeth meshed so that driving one of them with the handle turns the other, making the whole mechanism work as one.

If you look more closely, you'll see that as you turn the handle clockwise, the blade on the other arm rotates counterclockwise. The two gears are the same size, so they rotate at the same speed, and both complete a full rotation at the same time. Next, imagine you've deconstructed the can opener and laid the two gears side by

side—but now, the gear on the right is bigger than the one on the
left, with twice the circumference and twice the number of teeth.
Because the outer edges of both gears are meshed, they have to
move the same distance, so the smaller gear will complete two circles for every one completed by the larger gear. This also changes
the amount of torque acting on the gears; the smaller gear has a
bigger torque acting at its rim than the larger one, a fact that is
used when designing some well-known machines.

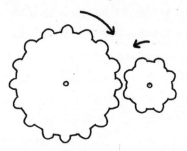

Meshed gears of different sizes

For most of us, our first experience of controlling gears is on a
bicycle. I'm no cyclist myself, but I have always watched professionals in awe as they navigate curves and slopes, deftly switching
their gears as they build up and control their speed. Their pedals
are attached to gears (known as the chainring), which are in turn
attached via a chain to a set of gears of different sizes on the back
wheel. A modern bike typically has up to three gears at the front
and eight to eleven at the back, but for simplicity, let's imagine
there is just one front gear, and it has forty-eight teeth. Cycling on
a flat surface is easy, so we engage a small rear gear: say, one with
twelve teeth. For every pedal stroke, the larger front gear turns
the small gear (and the back wheel) four times (forty-eight divided

by twelve), so we travel the farthest distance for one rotation of the pedal. But when we're riding up a steep hill, lifting up our weight against gravity, the amount of force needed to turn this small gear becomes too much for our legs. We switch to a larger rear gear (say, one with forty-eight teeth). It becomes easier to pedal as the force reduces, but we travel a shorter distance. Now the rear wheel only turns once for every cycle of the pedal.

Gears allow us to change the direction and magnitude of forces. Over time, this has enabled engineers to develop a whole suite of technologies. The steam-powered train could, through a system of gears, use fuel in its engine to turn wheels on its carriage. In a watch, different-sized gears, all powered by the same source, send the second-, minute-, and hour-hands around the dial at different speeds, making our reading of time more accurate. We change gears in our cars to help us climb steep slopes, to consume less gasoline at higher speeds, and to stop us from rolling down a hill uncontrolled. Factories have large, building-sized manufacturing lines that rely on gears to make all sorts of stuff. However, while it's easy to be impressed by the way gears power large and complex items or the impossibly small and intricate, some of my favorite examples of gears at work are in everyday household appliances— blenders, washing machines, dishwashers. These might seem less glamorous than Rolexes and racing bikes, but to me they're anything but, because of how they have transformed society in general, and women's lives in particular.

In 1893, at an industrial fair in Chicago, one new invention was attracting the attention of the crowds—a large, rectangular wooden box, with a slew of cranks, spinning gears, and wheels on one side, into which a cage full of dirty dishes disappeared, only to reappear minutes later, clean, as if washed by hand. The exposition judges awarded it the highest prize for "best mechanical construction, durability, and adaptation to its line of work." It was the only machine on display designed by a woman.

Josephine Cochran was born in 1839 in Ashtabula County, Ohio, to a family of engineers. Her maternal grandfather, John Fitch, invented the first patented steamboat in the United States, and her father, John Garis, was a civil engineer who constructed a number of mills along the Ohio River. Had Josephine been born Joseph, perhaps there would have been the opportunity to follow in the family footsteps and study engineering, but this was an era with limited options for women. She married William Cochran when she was nineteen and moved to a mansion in Shelbyville, Illinois, where she became a socialite and a mother to two children.

The Cochrans enjoyed hosting large dinner parties, and Josephine often brought out her prized heirloom dishes, which were believed to date back to the 1600s. She became frustrated when they got chipped by the staff washing them. Sometimes she washed them herself to keep them safe, but she was convinced that there must be a better way of doing things. Perhaps her childhood observations of the engineers in her family inspired her to think big, and she began to design a machine that would do the job, saying, "If nobody else is going to invent a dishwashing machine, I'll do it myself."

In 1883, partway through this process, Josephine's husband died, leaving her with considerable debts and very little money. She was forced to quickly turn her ideas and sketches into a real business to live off. She sought help from professional engineers to assist her with the intricacies of the mechanics but was frustrated by their contributions, saying later, "I couldn't get men to do the things I wanted in my way until they had tried and failed in their own." She filed her first patent in 1885, and hired a young mechanic called George Butters to help her make the first prototype of her dishwasher in a shed behind her house.

Men—who I'm guessing didn't have much to do with washing dishes in that era—had attempted dishwasher designs before Josephine, but their designs had resulted in breakages. They used scrubbers to clean the dishes, and needed someone to manually pour boiling water over them. Josephine, on the other hand, understood how the dishes needed to be cleaned and protected, and she also knew that to be truly useful, the machine should require minimal input. She started her model from the inside out, measuring her dishes and designing a wire cage with specific sections that securely held different sorts of dishes. An arrangement of gears slowly rotated this cage so that all the dishes were exposed to soap and squirting water—her machine was the first one to use pumped water rather than scrubbers for cleaning. Each of these pumps (one for water and another for soap) was operated by a lever, which engaged a variation of the gear (rather than being circular, her "gears" were toothed arcs, and there were two pairs of them). As the lever was swung back and forth by its operator in its initial configuration, soap suds were released from the first tank onto the dishes. Then the arcs were adjusted into their second position to operate the pump that pressurized the hot water to rinse the dishes until they were clean. This whole process took only minutes. On December 28, 1886, Josephine Cochran was granted a patent for her dishwashing machine.

At first, because of the price of her machine, its size, and the amount of hot water it demanded in use, Josephine had difficulty selling to the domestic market. So, she focused instead on restaurants and hotels, completing her first order for Chicago's Palmer House, a famous hotel, thanks to a friend who made an introduction. She then wanted to sell to another large hotel, but without someone to help her, she had to go alone at a time when women of her social standing didn't leave home without a man. Having never been anywhere without her father or husband, she said that the

J. G. COCHRAN.
DISH WASHING MACHINE.

No. 355,139.　　　　　　　　　Patented Dec. 28, 1886.

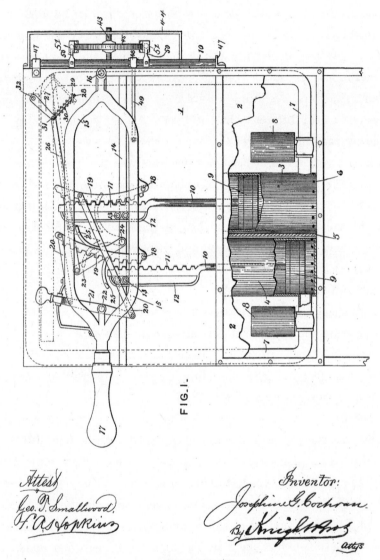

FIG. I.

Attest　　　　　　　　　　　　　　　Inventor:
Geo. T. Smallwood.　　　　　　　　Josephine G. Cochran
F. A. Hopkins　　　　　　　　　　By Knight Bros
　　　　　　　　　　　　　　　　　　　　　　　attys

The Cochran dishwasher, showing its modified version of gears

lobby seemed a mile long, but she crossed it, met the team, and left with an $800 order.

Although Josephine was now in business, she struggled to raise capital because investors weren't open to giving their money to a woman. She decided to press on regardless, and the 1893 industrial fair in Chicago provided the breakthrough she needed. The poster she displayed at the fair advertising the Garis-Cochran Dish-Washing Machine Company proclaimed: "Every enterprising and progressive hotel man should investigate this machine, it is simple and easy to run. A man can learn how to use it in an hour." Not only did she receive orders from establishments in Illinois and neighboring states in the US, but also from places as far away as Mexico. Her larger model could wash and dry 240 dishes in just two minutes, and her customer base expanded to include hospitals and colleges as well.

Josephine continued to develop her designs, creating a motorized model that featured racks that moved back and forth with mechanically pumped water. A later model featured more wheel-inspired rotation with revolving racks, and also drained water into a sink through a hose. In 1912, when she was seventy-three, the woman who was once terrified of crossing a hotel lobby alone went to New York and sold multiple machines to prestigious hotels and shopping centers. Sadly, she died just the following year, and her business was bought by the Hobart Manufacturing Company, and amalgamated into their KitchenAid division, now part of the Whirlpool group. KitchenAid introduced the first successful domestic dishwasher in the 1940s, finally bringing the machine to Josephine's intended audience.

Today's dishwashers still use pumped hot water to clean dishes, just as in the pioneering Garis-Cochran machine. In another nod to the wheel, they also have spinning arms that spray hot water over the dishes like a sprinkler system for a lawn. This was an

innovation that Josephine's engineering genius had already antic-
ipated; just before she died, she came up with a design along
these lines.

In 2006, Josephine Cochran finally achieved recognition for
her inventions by being posthumously inducted into the American
National Inventors Hall of Fame. Inventions like the dishwasher,
vacuum cleaner, blender, and washing machine have changed millions of lives. All of them are based around spinning components
likes gears, wheels, and pulleys, which make them compact enough
to fit in a home. Before these devices became widely used, women
spent several hours a day on chores. Once they could use machines
instead, they gained something incredibly precious: time. A University of Montreal study published in 2009 says that in the year
1900, women spent an average of fifty-eight hours per week on
household chores; in 1975, it was eighteen hours. Appliances were,
of course, a small part of a bigger picture of other developments,
such as increasing wages, wars, and broader discussions about a
woman's role in society; nonetheless, household appliances played a
role in women making space for themselves in the workplace.

The different forms of the wheel used in carts, cars, bicycles, and
trains completely changed the landscape of our transport on land,
and how we designed our cities around them. Chipping away at its
circumference created a whole new world of machinery. Turning
the wheel back to the same orientation as the potter's wheel and
attaching blades to it gave us helicopters, and the engines of our
aircraft couldn't exist without wondrous whirling wheels, opening
the sky to humans, and making our world a much more accessible
and dynamic place. But all this progress was the result of the reinvention of the wheel itself. For some of the most groundbreaking
developments in navigation, engineers had to turn their attention
in a different direction.

Something amazing happens when you allow not just the wheel to spin but the axle, too. Doing so enables us to orient ourselves in extreme environments, and opens up exploration into the far beyond. This miraculous machine is the gyroscope. At its heart is a spinning wheel, but this wheel is suspended within a frame that has a series of rotating arcs, which means the wheel can spin in any direction it wants. The reason this system is useful in creating stability is because it harnesses a special behavior of moving objects.

Newton's First Law tells us that if an object is moving uniformly, it will continue to do so unless some external force causes it to change. The moving object that Newton describes has momentum—which is a measurement of the object's mass, speed, and direction of travel. So, a pool ball rolling across a table will continue to do so until it hits the edge of the table.

This momentum is *conserved*, meaning the total momentum of a system must always remain the same. Picture two pool balls traveling directly toward each other at exactly the same speed. Because they have the same mass and speed, but are traveling in opposite directions, their total momentum is zero. After they experience a perfect collision, they immediately start moving away from each other with the same speed, still with a total momentum of zero. The pool balls are an example of linear momentum, which is the momentum of objects moving in a line, but this principle of conservation also applies to the momentum of spinning objects, which is known as angular momentum.

Newton also told us that for every action, there is an equal and opposite reaction. If I push against a wall, the wall pushes back at me. This also applies to rotational force, which is called torque. Bringing all this science together tells us that spinning objects have momentum; if you try and change their orientation by applying a force, they push back, and the total angular momentum of a system always remains the same.

The idea behind a gyroscope is to set its rotor—called the flywheel—spinning in a particular orientation. Because it has to conserve its angular momentum, even if the frame rotates around it, the wheel will remain in its original orientation. The axle of the wheel doesn't move when the frame around it moves, because it's made from gimbals. On some airlines (in economy class; I can't speak to anything else), you often have a little holder just for your drink, which you can unclip while the main tray is still attached to the seat in front of you. The holder has a ring within it, which is pinned at two points, allowing it to rotate, so you can place your cup in it and adjust it so it's vertical, regardless of the angle of the holder—this is a gimbal. The gyroscope's frame is made up of two gimbals. The axis of the spinning wheel is attached at each end to the first gimbal, and this gimbal is, in turn, attached to the second one, also at two points, but at ninety degrees from the wheel's axis.

Gyroscopes have had a huge impact on travel and exploration. Before they were invented in the nineteenth century, navigators were dependent on the magnetic compass, which, in turn, depends on the Earth's magnetic field. This, however, is far from fail-safe, because the Earth's magnetic field isn't fixed; it moves. This means

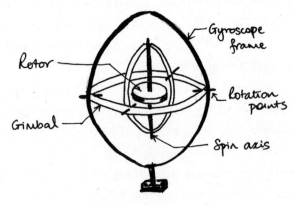

Example of a gyroscope

that "north" on the magnetic field doesn't always coincide with north as measured by the Earth's rotation. In addition, metals could interfere with the compass, as sailors realized when they first sailed out to sea in iron ships. The gyroscopic compass avoids such problems. Even as the gimbals rotate, the flywheel maintains the orientation of its axis. Comparing the position of the axis relative to the frame gives us reliable information about which way we're pointing. A fighter pilot putting her plane through a series of intricate twists and turns at speed can still be sure of how she is oriented, thanks to gyroscopes.

Gyroscopes can also be used in the opposite way. Rather than allowing the flywheel to merrily spin away, you can force it to change direction by reorienting its axle. Take a look at the children's toy, the top. A spinning top's angular momentum points upward through its axis. If you have two identical tops that are spinning next to each other at the same speed, together they have twice the angular momentum (this is analogous to two pool balls traveling at the same speed, in the same direction, in parallel lines). If you are in space, and the tops are floating so that one is upright and the other is upside down relative to each other, then the total angular momentum of the system is zero (which is the equivalent of the two pool balls traveling directly toward each other).

Let's take this out of the playroom and into the wider world. Imagine a structure in which you have four identical flywheels that are mounted in gimbals—so, four gyroscopes. But this time, rather than allowing the gimbals to move freely around the flywheel, you can control the gimbals and therefore the direction in which the axles of these wheels point. Each flywheel weighs 98 kilograms, is about 1 meter in diameter, and spins 6,600 times every minute, so they're relatively large and extremely rapid. They sit within a much larger structure. Normally, they are oriented such that their angular momentum adds up to zero, so they don't have any effect on the outer structure. But if you need to, you can tweak the gimbals and move the wheels so that the total momentum increases in value, going up all the way to a maximum when they are all pointed in the same direction. The principle of conservation of angular momentum tells us that because we started off with zero momentum, but now have forced a net momentum out of the gyros, the gyros will push back, causing the larger structure within which the gyro sits to rotate to cancel out this momentum back to zero. This larger structure just happens to be the International Space Station (ISS).

The ISS must be the most impressive example of human collaboration. Multiple teams from Europe, Japan, Canada, Russia, and the United States have been working together for decades to create this space station, which allows scientists and engineers from all over the world to work on experiments without gravity. The station (which has been gradually constructed 408 km above the Earth's surface by joining together modules) weighs 419,700 kg and is 109 meters long by 73 meters wide, slightly larger than an American football field. The ISS has been given a very specific orientation, called its "attitude," so that its orbit is at a fifty-one-degree inclination to the equator and takes ninety minutes to complete. This path covers over 90 percent of the planet's population, giving the station a unique vantage point of the Earth's surface. The ISS is also set so it always has the same side facing the Earth. The main reasons for this are because if the same side of the station always faced the Sun, then it would get heated to extreme temperatures while the other side would be extremely cold, affecting the materials from which it's made, and also because the communication satellites the ISS uses to talk to Earth are up in high orbit, so the antennae always need to point away from the Earth to maintain communication.

In space, as movies and sci-fi shows have shown us, once you set an object spinning, it continues to spin unless an external force changes it. In principle, this means that the ISS could be set in its orbit and then left forever, but tiny forces acting on the station influence its attitude. Even that far away from the Earth, there is a little bit of atmosphere, which creates a small resistance to the station's movement. Its main power source—solar arrays that are mounted on large adjustable wings—are moved around to face the Sun, and if they are in certain positions, the small air resistance increases. There is also a minuscule effect of gravity gradient: the farther away something is from the Earth, the smaller the pull of gravity, so the parts of the ISS closest to the planet experience a slightly larger pull than other parts, which creates tiny out-of-balance forces.

The ISS, therefore, needs ways of reorientating itself, and gyroscopes play a part in this. Sci-fi has shown us that using boosters or thrusters is one way of making directional adjustments. But while there are thrusters on the ISS for larger movements (such as when the ISS needs to change attitude significantly to receive an incoming docking vehicle), they're not always the best option. For a start, they need precious fuel, which has to be transported from Earth, an extremely expensive thing to do. On top of this, the force from the thrusters can be large, making it complex to control if subtle shifts are needed. And these forces can also affect the extremely delicate experiments that the astronauts are conducting in the microgravity environment of the spacecraft. The solar arrays that power the ISS, for example, are fragile and generally need to be parked and locked if the thrusters are engaged, which is pretty disruptive to the operation of the ISS. And so, to instigate small and gentle shifts to pilot the structure, engineers use the four gyroscopes, which are called Control Moment Gyros (CMGs).

In normal circumstances, the ISS more or less flies itself—a system of computers and sensors measure and record its attitude, and feed information to the computers controlling the CMGs, telling them if they need to be moved, and by how much. This system is so sensitive that it knows when the crew are waking up and starting to move around the station, which creates tiny shifts in the angular momentum of the system as a whole. The team that pilot the ISS from the ground spend much of their time running simulations of scenarios to understand what should be done if something goes wrong, and planning for unusual activities, such as a docking module. They hold primary responsibility for controlling the attitude, freeing up the astronauts to get on with their busy schedules. But if a CMG starts misbehaving, for example vibrating within its bolt-, spring-, and magnet-laden casings, then the team in Houston can take manual control.

The wheel, then, inspires and controls technologies from the minute to the immense. It is both bewildering and breathtaking that, 408 km above our planet, there is a laboratory in which astronauts are exploring whether microbes like bacteria and fungi can mine minerals in space, investigating how humans perceive time during long space flights, developing 3D bits of organs from stem cells to create artificial organs, testing new materials and growing food— not only with a view of improving technology here on Earth, but also to imagine the possibility of humans living on extraterrestrial bodies. And all that happens because humans are able to sit at a desk on Earth and speak to satellites that are halfway to the Moon, in order to control four wheels that keep that laboratory stable.

Since its creation, the wheel and axle have allowed us to create machines that have impacted migration, language, and globalization. Now they are helping us imagine a future beyond the confines of our planet. Human progress and the reincarnations of the wheel and axle are intricately intertwined. And that's why we should absolutely continue to reinvent the wheel.

Spring

If we bounce back in time, to the childhood of Chinggis Khaan (or Temujin, as he was then called) in the twelfth century, we would doubt his chances of growing up to unite the disparate tribes of Mongolia and establish the largest contiguous empire in history. After his father was killed by a rival clan, Temujin, his mother, Hoelun, and his siblings were abandoned. Their tribe disappeared one night with the family's animals, just as winter began to set in: a winter that would see temperatures plummet to forty degrees below freezing. No one expected—or wanted—them to live. That they did is largely due to Hoelun, who taught her family the necessary skills for survival: foraging, stitching together animal skins for warm clothing, hunting with bows and arrows. As Temujin became Chinggis, this last skill made the difference between survival and supremacy—a key feature of his military tactics would be an exceptionally light and powerful bow.

The bow is a type of spring. Fundamentally, springs are things that store energy when their shape is changed by a force. When the force is removed, they ping back to their original shape, and

release that energy—and this energy is made to do something useful. They are made from semi-flexible materials, so it's relatively
easy to deform them, and they are elastic, meaning that the materials regain their original shape after the force is gone (like a rubber band being pulled and released). Plastic materials, on the other
hand, get permanently deformed by forces: poking modeling clay
with your finger creates an imprint that lasts.

Deforming a spring takes energy. An archer holds the curved
bow in front of her in one arm and uses the string that is tied
to each end of the bow to change the bow's shape as she draws
her hand back. The bent bow stores this energy until she flicks
open her fingers, at which point the energy is rapidly released into
the arrow, which shoots off into the distance. The more the bow is
deformed, the more energy it stores, and the more powerful it is as
a spring; throwing an arrow with an arm simply can't compare.
The spring was humanity's first tool that allowed us to *store* energy
and then release it *when we wanted*, often amplifying our effort.

I find springs particularly interesting because of the huge diversity of their forms. Unlike the nail and the wheel—whose silhou-
ettes have hardly altered since they were invented—springs come
in myriad shapes. A spring can be bow-shaped, like the weapon
of course, but also like the springs made from layers of slightly
curved wood or metal, which formed suspension systems for carriages. It can be a cylindrical helix—like the familiar coiled metal
ones, only patented in 1763—but the coils can also be spiral, con-
ical, or spherical. A spring can even be as simple as a cube- or
cuboid-shaped block, molded from layers of rubber, a form that's
more familiar from my work in structures.

. The way they can be forced and deformed also varies. They can
be configured to store energy when squashed: a retractable ball-
point pen has a long, narrow spring just inside the tip that gets
pushed down and held in place while we're writing. When we're

done, we flick the spring free, which lengthens to make the ball-point disappear. Springs can also do the opposite, holding tension when stretched, like those that line the edge of trampolines. When you jump down, putting your weight onto the net, the springs around the edges lengthen to store energy, which is released as they shorten, back into the net and into you, sending you much higher than is possible with just your legs. Springs can even work when twisted, in torsion, like the little ones in clothes pins. And they can hold more power the more they are forced to curve, like the carriage's suspension spring, and the bow.

Because of this flexibility (literally and figuratively), springs are probably used in the widest range of stuff, and at the most varied scales, of all my selected inventions. They are found in weapons like bows, but also catapults and firearms. They form parts of wristwatches, pens, tweezers, keyboards—anything with a push button. We find them in locks, mattresses, stretchy gym equipment, sprung dance floors, trampolines. They play a role in suspension systems for strollers, wagons, and cars, and also retractable needles in injections, and stable bases for microscopes and telescopes. Even the foundations of some skyscrapers have giant springs that help stabilize them against the shaking of earthquakes. Springs, for me, are the epitome of versatile engineering.

As to who invented the spring, and where, and when, it's hard to say. It's likely that the bow and arrow was one of its earliest, practical forms that was complex—making and stringing a bow, then aiming and releasing an arrow, was a mark of increased ability in humans. But because bows were made from wood, an organic material prone to degradation, they didn't survive the many tens of thousands of years we think they've been in existence. The oldest complete bow that has been excavated dates back to only 10,000 years ago. Broken into five pieces, this deep brown elm-wood bow is known as the Holmegaard bow after the area in Denmark where it was discovered in the peat bogs during the Second World War. Like the English longbow, a famous example from medieval times that is known for its success against the French in the Hundred Years' War, the Holmegaard bow is made from a single material: wood.

The reason medieval Mongolian bows were so remarkable is because they were the most advanced composite design of their time, meaning they were made from multiple materials. Bow designers understood that when drawn, the inside face (closest to the archer) is squashed or compressed, while the outer face is stretched, or in tension, and they used the strongest materials for the type of force in its construction.

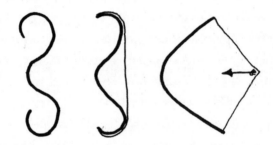

Simplified diagram of a Mongolian bow with its curvature

The core or central portion of the bow was made from bamboo or wood, a light but strong material, which was carefully dried so it was robust and flexible. Then, the tendons or sinew from the hind legs of a deer, cow, horse, or moose were dried and beaten to create loose fibers. These fibers were dipped in a glue made from fish air bladders and layered on the outer face of the bow, which was a painstaking process: too many fibers would make the bow too stiff and harder to deform, meaning less energy for the arrow; too few fibers would make the bow weak and prone to snapping. Sinew is a perfect material for bow-making, because it's flexible and strong when stretched. This is because of collagen, which most of us know mainly from anti-wrinkle face-cream ads and media speculation about which celebrities have had injections of it, but is in fact the protein molecule that gives connective tissues like sinew their strength.

After six months of drying, horns from wild goats or domestic animals were boiled to soften them, processed, and then stuck to the inside face of the bow to resist the compression. Another six months of drying, and then the bow was protected against moisture with the application of a layer of softened bark. The resulting weapon was weather-resistant, strong, light, small, springy, and very powerful. Even though the English longbow from a similar era is larger and may seem more sinister, the flexibility of the Mongolian bows meant that without arduous effort from the archer, they stored and released more energy, meaning faster and farther-reaching arrows. A stone stele (an upright slab), believed to date back to 1226, records in the ancient Uigarjin script that, during an event to celebrate the conquest of Sartaul in East Turkestan, Esungge the Marksman shot a target at 335 alds (which translates to around 536 meters). If accurate, this is remarkable compared to the English longbow, which was believed to average around 300 meters.

The Mongolian army under Chinggis Khaan was extremely effective, because soldiers used these bows while skillfully maneuvering hardy, small, and nimble horses. The soldiers had learned to ride at three, and were taught archery while riding with just their legs from the age of five. In battle, this meant their arms were free to shoot accurately at enemies from distances at which they themselves could not be harmed.

During his expansion, Chinggis Khaan also set his sights on China. He arrived at the capital, Yengking (now Beijing), which was designed to be impregnable: it was surrounded by thick, tall walls with close to a thousand guard towers, each defended by enormous crossbows. The Chinese also had trebuchets or catapults—weapons that also harness the physics of springs, and which launched clay pots filled with boiling petrol. A direct attack seemed an insurmountable challenge, so instead, the Mongols devastated the surrounding countryside on horseback until the people of Yengking starved and surrendered in 1215.

When Khaan left China, he took these bulkier, spring-driven Chinese weapons with him. The crossbow is a more sophisticated version of the bow and arrow, made up of a lock and a stock (I'll get to a couple of smoking barrels later). Instead of a person's arm holding the middle of a bow, the crossbow has a solid piece, usually made from wood, called the stock. The string is drawn back manually and held back by a lock, with an arrow placed in front of it. Activating the lock's trigger releases the string and arrow.

While the crossbow is still a spring, utilizing the same forces of tension and compression, it's different in many ways to the longbow. I tried traditional archery once. The bow I was given was very light, so I was surprised at how many of my muscles were needed simply to draw the string and hold it in place—my arms, yes, but also my chest, core, and legs. With all these muscles taut, it was challenging to hold steady, let alone aim and release the arrow (although nowhere near the bull's-eye, I managed to hit the board, which was victory enough for me). The crossbow, on the other hand, didn't require much training and muscle development, so peasants could be quickly recruited to supplement an army, and also defend themselves against professional soldiers. Some crossbows had far greater power, too, allowing arrows to fly with more force, so they could penetrate metal shields and armor. And, since the string is held by the lock, you could focus on your aim, making it much easier to fire accurately.

Over the years, the Chinese designed different variations: smaller crossbows that fired poisoned arrows with just one hand,

models that released multiple arrows, and heavy artillery versions that were mounted on movable bases. Increasingly, crossbows required less skill, less training (and thought) to operate,
and could cause death and destruction from a distance. Before
Chinggis Khaan arrived, the Chinese had succeeded in withstanding invading armies in part through their skill in making and
deploying crossbows. Despite their weaponry, Khaan managed to
overcome them because of the agility and ability of his skilled and
self-sufficient army to quickly traverse long distances with deadly,
highly portable Mongolian bows.

This, no doubt, contributes to Khaan's reputation as a barbaric
killer. In fact, his legacy is more variegated than that. Khaan
allowed his subjects to practice their own religions (which was
extremely rare at the time), and he promoted people, even from
places he had defeated, if he felt they were talented. The thirteenth and fourteenth centuries were an era of relative peace and
prosperity in Eurasia. The Mongols established a single system of
trade and tariffs, and created the *Yam*, a postal service that connected lands as far as eastern China to Syria. Traveling became
safe (assuming you'd surrendered), increasing trade along the Silk
Road, which in turn spurred the cross-cultural exchange of technologies and goods, bringing, among other things, silk and paper-
making to medieval Europe.

It's tempting for me to paint a picture of engineering as a force for
advancement and good in the world, of course, but that's only a partial perspective. Trade routes also introduced to Europe the bow and
arrow, the crossbow, and the catapult—weapons that were deadlier
than those that came before them, thanks to the science of springs.

Springs still play a central role in weaponry (now for the smoking barrels I promised earlier). Modern machine guns can fire multiple bullets in succession without reloading, usually because of an
arrangement of springs driving the belt of bullets through the bar-

rel. Semiautomatic pistols have multiple springs as well. In such a pistol, there is a removable unit within the handle, called the magazine, which is where the cartridges are stored. A large spring at the base of the magazine gets compressed as you load up the bullets. Once the magazine is put back inside the handle, you pull back the top portion of the gun, called the slider. The slider has a spring called a recoil spring, which gets compressed. Moving the slider creates an empty space above the magazine, and the spring at its base pushes up a cartridge. As you release the slider, the recoil spring pushes it back into place, and the gun is loaded and ready to fire. Pulling the trigger releases yet another spring, which is connected to the firing pin—a long, sharp rod of metal—which hits the cartridge sharply. This causes a small explosion in the cartridge that blasts the bullet out. Although it's taken me a paragraph to explain this sequence of steps, springs ensure that all this happens within a tiny fraction of a second. But their work isn't done yet. Thanks to Newton's law of equal and opposite reactions, the explosive force of the bullet firing compresses the recoil spring, pushing the slider of the gun backward, which allows the next cartridge to pop up, ready to be fired.

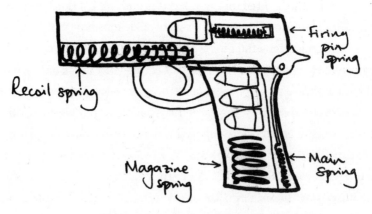

Springs in a semiautomatic pistol

The story of the modern gun has links to the bow and arrow. The Chinese invented gunpowder—a mixture of charcoal, potassium nitrate, and sulfur—in the ninth century and used it to fire flames and shrapnel at their enemies from hollow bamboo or metal tubes. During the Mongolian era of trade through the Silk Road— which I'm conjecturing was stabilized, at least in part, due to the bow—gunpowder arrived in the hands of the Europeans in the thirteenth century. By the sixteenth century, they had created firearms that were far more advanced than their Eastern predecessors. Intimidation and conquest then became about firepower, and once again the face of war evolved. The distance between dueling armies became vast, and the type of injuries sustained changed, which necessitated advancements in medicine. Until the devastating atomic bomb that could obliterate large sections of a country, war was dominated by firearms.

The power of the gun scares me. With minimal force from just one finger, someone can die, because the science of springs has been so effectively harnessed. Gun violence is one of the biggest public-health crises of our time, according to Dr. Mohsen Naghavi, the lead author of a 2018 report by the Institute for Health Metrics and Evaluation. In 2020 and 2021, over 45,000 people were killed each year in the United States. Apart from 1994, the year of the Rwandan genocide, global firearm-related deaths have been higher than global conflict and terrorism deaths every year since 1990. These statistics are humbling, particularly for an engineer. I think it is important that we remember our inventions and innovations can be deployed for the destruction of human life, even if that was never our intention.

Springs are an interesting example of a piece of engineering that was used over a long period of time in many forms before the science behind it was properly studied. The makers of springs operated by trial and error, and used their experience to figure out what materials and shapes worked best to suit their purpose. It's a testament to the progress we can make without understanding the details of how something works. Nevertheless, once the science *was* studied in the seventeenth century, our use of springs became far more sophisticated.

This jump in understanding and application is largely due to Robert Hooke. Hooke was born in 1635 on the Isle of Wight in England. The youngest of four children, he showed an interest in machines and drawing from an early age, and went on to study at Oxford and make a name for himself as a true polymath; he built telescopes and observed the movements of Mars and Jupiter, hypothesized about fossils and the age of the Earth at a time of literal biblical interpretation, and surveyed around half of the buildings damaged in London's Great Fire of 1666. One of his many interests was the elasticity of materials, which led him to study springs.

With the aim of mathematically and physically explaining how springs work in a predictable manner so that they could be used in mechanisms, he published his thesis, now known as Hooke's Law. Oddly, he presented this cryptically as an anagram in 1660, providing its solution eighteen years later as *"Ut tensio, sic vis,"* which is Latin for "As the extension, so the force." What Hooke explained is that the extension of a spring is proportional to the weight it carries, and the more a spring is extended, the larger the force needed to deform it, and the more energy it stores. In other words, if a weight of 1 kg causes a spring to extend by 1cm, then a weight of 2 kg causes a 2 cm extension—and in the latter case, the spring holds more energy (this law quantified why Mongolian bows were so effective). Hooke's Law holds as long as the forces and deformations are small enough so the material stays elastic (i.e., it can regain its original shape).

This law was far-reaching. Beyond springs, Hooke's Law allowed engineers to predict how much an elastic system would expand, contract, or move under particular forces. This could be how high a ball might bounce, how a material might absorb sound, or even how much a tower might sway under wind and earthquake forces. Hooke's work also had many practical applications: it's obvious that a coiled spring for the suspension of a car is different from the springs used to enlarge constricted blood vessels in our bodies. Once you understand the application, including what the forces are and how much movement or flexibility you need, you can design the spring. Assuming it's a coil, you can work out its overall diameter, the diameter of the wire, its length, and the number of coils, ensuring they are perfectly balanced. Using these principles, engineers designed devices like the spring scale (my daughter was suspended in one of these a day after her birth to check her weight accurately), pressure gauges (like the old-school blood-pressure checkers where a needle spun around a dial as your arm got squeezed and released), and complex locks that are difficult to pick (small springs fall into place on the ridges of keys and, when they're in the right configuration, release the mechanism). I could have explored the role of the spring in any of these devices next, but instead I've chosen a mechanism that made a huge evolutionary leap thanks to the spring, and changed the way we live and work forever.

In one of the tall redbrick Georgian buildings in the Jewellery Quarter in Birmingham in central England, I visited Dr. Rebecca Struthers, a watchmaker and horologist. This would have been an interesting and instructive experience at any time, but, having been stuck at home for months riding out the various lockdowns in the 2020 pandemic, it was a real pleasure to visit her in her workshop.

It was pristine: a large, open space with high, white tables. Apart

from her dog, Archie, barking from an adjacent room, excited by
my arrival, the place was quiet. Around the room was a treasure
trove of old tools, like lathes, milling machines and topping tools—
each had a name, Rebecca told me, and was a part of her family.
Here, she and her husband, Craig, run their business, Struthers
Watchmakers, which specializes in making beautiful bespoke
watches and restoring historic timepieces—anything from creating
from scratch a new bezel for a Universal Genève to deep-cleaning
the dashboard clock on a vintage car. What with her fascination
for vintage watches, I was surprised to see a gold digital Casio on
Rebecca's wrist. But she explained that she needs a "workshop-
compatible" watch when she's working, one that won't get damaged
if it's knocked around as she moves between her precious machines.

Apart from being incredibly talented at her work, Rebecca is
unique in her industry. Being raised by a stay-at-home father while
her mother went to work in the eighties showed her the power of
smashing stereotypes, so that's exactly what she did in becoming
a watchmaker—a young, tattooed, working-class woman maker
(as she described herself as we chatted, face masks in place) in
an industry dominated by white middle- and upper-class men.
In 2017, she became the first watchmaker in British history to
earn a PhD in horology. Rebecca outlined to me some of the challenges people faced telling the time before timepieces with springs
were developed.

The key to measuring time scientifically and with some measure of accuracy is to count something that happens periodically
and regularly. Around 5,000 years ago, the ancients in Egypt and
Babylon created structured timekeeping based on three natural
phenomena: the daily movement of the Sun, the monthly cycle of
the Moon, and the annual change of seasons. The number twelve
had significance for the ancient Egyptians: at the annual flooding
of the Nile, a dozen special constellations could be seen, and they

divided the year into twelve months of thirty days (with an extra five days to complete the year). Each interval of light and darkness was split into twelve equal parts, creating a temporal hour. (Temporal hours varied in length as the length of day and night fluctuated, so in the summer, temporal hours were longer in the day than the night.)

This was a pretty good way of measuring time, but it had its limitations: because the ancients were reliant on sunlight, which wasn't always available; because of the variation in the length of days as the Earth moved across its orbit; and because the cycle of the rising and setting of the Sun isn't very frequent.

To physically see the passage of time, our ancestors used sticks to track shadows to give them an indication of how much of the day had passed. Later, water clocks were invented, which allowed them to make measurements day or night. In their simplest form, water clocks are basins with a small hole from which water drips into a container, indicating by its level how many hours have passed. Just as the stick evolved into sophisticated sundials, water clocks, too, became incredibly nuanced over hundreds of years, with the inclusion of complex arrays of gears. This development was led by Arab and Chinese scientists in the medieval era. Both these systems worked well in warm climates, but in northern Europe, cloudy days and freezing nights rendered them less useful, and the Europeans had to find another way.

Before the invention of electricity, clocks in Europe relied on mechanical systems to supply them with power. The earliest of these mechanical clocks appeared in Italy in the thirteenth century in Catholic churches, which might seem an unlikely place for innovative engineering until you think that members of the clergy were expected to pray at least seven times a day and once at midnight, and the priests needed some way to alert them to perform their duties.

Perched high at the top of church towers, these clocks had a
heavy weight that descended slowly down the inside of the tower
because of gravity. A series of mechanisms controlled how quickly
this weight moved and absorbed its energy to drive a chain of gears.
Part of this chain was a lever, and every hour, this lever would be
released to strike a bell.

As it was driven by the constant force of gravity, this clock's
source of power was reliable, but descending weights were not
exactly practical to dangle from your wrists. Jangling the weight
around would affect its reliability. When the large weight had
descended, it needed to be lifted back up to start its descent again.
In order to avoid having to do this multiple times a day, a long
drop was needed, hence the height at which these timepieces were
installed. So, clocks were large and immobile—until the invention
of the mainspring.

In Rebecca's workshop, she had laid out an array of springs on a
worktable for me to see. There were tiny springs shaped like question marks that fit on the tip of my finger, and marginally larger
V-shaped ones made from thin, cylindrical wire. Among them lay
mainsprings: spirals made from flat, wide ribbons of metal. One
was left loose on the table, free to take its natural form—tightly
wound at its center, but then opening out, like a Fibonacci curve,

Curled and uncurled mainsprings of different forms.
Image taken at Struthers Watchmakers

as large as the palm of my hand. Next to that was a similar spring whose end wound back the other way, making me think of the tail of a seahorse.

Another one had been wound up tight and placed inside a circular case called a barrel. These flat spiral springs are the source of power of small, mechanical clocks and timepieces: power that is needed to drive chains of gears that measure time. By replacing large moving weights, they enabled an evolutionary leap in timekeeping. Without them, watches small enough to fit on our wrists wouldn't have been invented.

Mainsprings are believed to have been invented in Germany toward the end of the fifteenth century. Flat ribbons of metal were formed into spirals and coiled around an axle. The inside end was attached to the axle. By winding it round, the spring tightened up and stored energy. To stop the spring from quickly unwinding, a device called a ratchet was used. This is a gear with specially shaped teeth that has a lever attached. When rotated in one direction, the lever allows the gear to move, but if rotated in the opposite direction, the lever catches the teeth and blocks the gear. (This is what makes the clicking sound when you turn the knob on a wind-up toy or music box.)

Once wound up, the axle is released, and the spring slowly unwinds, sending its energy into the gear train to drive the clock. Typically, the mainsprings were made of a length that meant they could last forty hours without needing to be wound up. Ideally, the wearer would do it once a day, but this provided a cushion in case they forgot.

Rebecca explained that replacing the large weight in clocks with the mainspring meant that clocks could now become smaller and even portable, and pocket watches were created for the first time. However, until Hooke figured out the science behind springs in the mid-1600s, working with springs could be a hit-and-miss affair; the first mainsprings made timepieces less accurate. This is because at their tightest, springs hold the most energy, and when unwound, they have less energy. Various mechanisms (with evocative names like "stackfreed" and "fusee") were added to the system to compensate for this, which, together with the mainspring, created a new, consistent, and compact source of energy.

Nowadays, stackfreeds and fusees are redundant because of the type of spring that has a tail like a seahorse, shaped so that it's stiffest at the tail end and least stiff at the start. In accordance with Hooke's Law, having a stiffer portion at the end means that, even though the spring has unwound, and therefore has less energy, its stiffness compensates. This puts a consistent amount of energy into the watch, so it doesn't lose time.

Next to the springs on Rebecca's worktable was an ornate nineteenth-century watch. I observed, fascinated, as she painstakingly deconstructed it to show me a second important spring: the hairspring. While mainsprings on their own made clocks smaller and portable, putting mainsprings and hairsprings *together* led to the creation of the first ever timepieces that were both accurate *and* portable, with levels of precision never seen before.

Once she had removed the lid by unscrewing some very, very small screws, Rebecca let me hold the watch. Terrified of dropping it, I nestled it carefully in my palm, squinting slightly to take in all the little mechanisms ticking away. The components were made of gilded brass. Intricate layers of differently sized gears gave off a dull, golden sheen. I could see the gears rotating at different rates depending on which hand on the clockface they drove—an impatiently speedy one for the second hand, which completed a full circle in a minute, in contrast to the languid pace of another that only completed a full turn in twelve hours.

But amid the more familiar gears was an unusual wheel, one that rapidly turned clockwise and then counterclockwise, pulsing

back and forth; this was called the balance wheel. The object that
kept this wheel oscillating quickly and consistently was a very fine
spiral spring: the all-important hairspring.

The Egyptians hit a wall on the accuracy of their time measurements because they relied on the oscillation of the Sun, which
was very slow at twenty-four hours. Accurate timekeeping needs
a form of oscillation, cycle, or back-and-forth motion that happens
consistently and frequently, so that large errors or inaccuracies
don't add up. Before the hairspring was invented, an ingenious
piece of engineering called an escapement was used in those large
medieval clock towers from the thirteenth century. Escapements
came in many forms, but they all worked on the principle of creating a regular swinging action to regulate the clock mechanism
in order to accurately record the passage of time. In contrast to a
twenty-four-hour Sun cycle, escapements had a cyclical time period
of just seconds.

One example of an early escapement is the verge and foliot,
which was common in the clock towers of medieval cathedrals.
It involved a rotating axle that was oriented vertically, with two
small tabs of metal attached to it. At the top of the axle were metal-

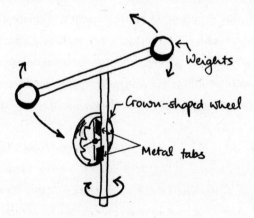

Verge and foliot escapement

lic arms extending sideways. Each arm held a lump of metal that could be moved to control the frequency of the oscillation. Every time the arm swung one way, the metal tabs twisted around with the axle, and released one tooth on a crown-shaped wheel. The tabs also forced the axle to then rotate the other way, and again release a tooth of the wheel. The crown-shaped wheel was connected to the slowly descending weight I mentioned earlier, so with each small turn of the wheel, the weight went down a bit. A cycle was thus created of the arms swinging back and forth, allowing the wheel to rotate a little, and in turn releasing the weight down slightly. The descending weight sent energy into the gears that operated a lever, which struck the bell of the clock to mark the hours. (The word "clock" originates from the Latin word for bell, *clocca*; the clocks didn't have a clock face or hands to tell the time.)

As crucial as escapements were in controlling how much energy the weight released into the system, the early versions were very sensitive and inaccurate. They were inaccurate because the oscillating mechanisms were inconsistent. For example, with the verge and foliot, friction between the moving parts, changes in temperature, or small movements in the position of the lumps affected the rate of the oscillation. When the frequency of the oscillation was affected, this had an effect on how much the weight descended, and therefore the time at which the bell gonged, meaning that these clocks could lose as much as a few hours of accuracy every week. Such inaccuracy was problematic and perplexing, as this seventeenth-century letter of complaint in the *Athenian Mercury* shows:

> I was in Covent Garden when the clock struck two, when I came
> to Somerset-house by that it wanted a quarter of two, when I
> came to St. Clements it was half an hour past two, when I came
> to St. Dunstan's it wanted a quarter of two, by Mr. Knib's Dyal

in Fleet-street it was just two, when I came to Ludgate it was
half an hour past one, when I came to Bow Church, it wanted a
quarter of two, by the Dyal near Stocks Market it was a quarter
past two, and when I came to the Royal Exchange it wanted a
quarter of two. . . .

This disgruntled clock-watcher would have been happy to know
that a big improvement in escapement design had already been
developed by the Dutch polymath Christiaan Huygens, who in
1656 incorporated a pendulum into the design of clocks. This is a
weight attached to the end of a long, thin rod, which swings back
and forth very consistently. An anchor-shaped escapement at the
top of the pendulum rocked side to side with each oscillation and
released a tooth of a gear each time, thus driving the clock. Unlike
time, the evolution of technology is rarely linear, and many pendulum clocks were still being powered by a descending weight, as you
can see in old grandfather clocks, while in others the mainspring
was doing the job.

While this greatly improved the accuracy of clocks, they were still
not portable, in part because they were unwieldy, but also because
the slightest movement disturbed the pendulum and escapement,
creating errors in the timekeeping. So, they were of no use to people
exploring the world around them, whether locally or on ships across
continents. The invention of the hairspring changed this.

Together with the balance wheel, the hairspring forms a system that oscillates at a high frequency (making it accurate and
precise), and isn't affected when jostled around in our pockets, on
our wrists, or out at sea. As I looked at the watch that Rebecca had
put in my palm, I was mesmerized by the twisting of the balance
wheel. It was almost too quick to see, but each time it changed
direction, it hit a small mechanism called a pallet fork and pushed

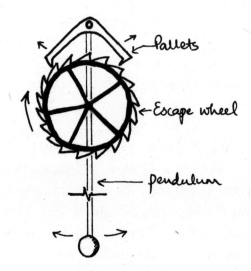

Pendulum escapement

it from side to side. The fork had two prongs that released one tooth of a specially shaped gear, called the escape wheel, each time it oscillated. (The soothing ticking you hear in a mechanical watch is partly this knocking back-and-forth action.)

The hairspring, with its associated mechanisms, forms the consistent, beating heart of the mechanical watch. (The name comes from the fact that the spring was originally made from a hog's hair. This wasn't a particularly reliable material, so it was eventually superseded by metal, but the name stuck.) It's an incredibly delicate thing; tiny changes in the length of the hairspring can slow down or speed up a watch, and creating the perfect balance is the mark of a great watchmaker. A timepiece owes its accuracy to its rate of oscillation: the more oscillations per second, the smaller the errors that add up, even when the watch is moved around, so the better the timekeeping. Whereas the verge and foliot could swing every few seconds, and the pendulum once or twice a second, the balance wheel and hairspring can oscillate multiple times a sec-

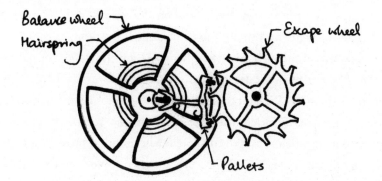

Balance wheel and hairspring escapement

ond. A typical mechanism will go six times a second, but there are some designs that oscillate more than one hundred times a second.

The advent of portable, precise pieces of engineering changed the way we lived. When they first appeared, these fashionable new timepieces were worn as pendants around the neck, suspended from waistbands, or carried in pockets; wristwatches were considered women's jewelry for a time. We can debate whether this was a good or bad thing—we now seem utterly beholden to counting the minutes down to our next task on a seemingly endless loop—but watches with springs had clear and profound impacts on science, shipping, and society.

Scientists could now observe celestial bodies like the Sun, the Moon, and stars, and record their position at a particular time accurately, then track their movement over the days and years, giving us a better understanding of our solar system and the universe beyond. Closer to home, before the railroads, each town set their own local time by observing the Sun. But in the nineteenth century, astronomers sent the accurate time across the country using the telegraph, ensuring that clocks were consistent along the whole network, not least to prevent collisions between trains that weren't at the right place at the right time. Timekeeping also had an impact on regulating trade, and on the length of days spent by workers in factories during the Industrial Revolution.

Keeping accurate time was also vital for the safety of sailors. After the Scilly naval disaster in 1707, in which around 1,400 people lost their lives, in large part due to the inability of the navigators to accurately calculate their position, the British Parliament offered a significant award to anyone who could improve the safety of long-distance sea travel. The problem lay in plotting the Earth's longitudes, a vital means of orientation when approaching land. (Polynesian navigators had been calculating longitude for years through natural observation and knew their patch of ocean well, but these techniques weren't used in the West.) The solution, at least in theory, was simple. If a clock was synchronized with the exact time at the zero-degree longitude at Greenwich, and then taken on a long-haul voyage, navigators could then calculate the time of their local position using the Sun and stars and compare this to Greenwich Mean Time. The difference in time would tell them how many degrees east or west they had traveled. This process, though, was dependent on an accurate clock. If it lost time unpredictably during the voyage, the calculation could be completely off.

John Harrison, a carpenter by trade and a self-taught clockmaker, spent decades working on designs of timepieces in the eighteenth century. He had already created the most accurate clocks on land, but clocks at sea presented a different challenge: they had to stay stable in the face of changes in temperature, humidity, and pressure; resist corrosion from salty air; remain accurate over very long time periods; and not be affected by the constant swaying of the ship. No timepiece had been created that could do all this.

In a demonstration of true commitment, over four decades Harrison designed and made five timepieces, known as H1 to H5. The first three of these are clocks, but H4, completed in 1759, looks more like an oversized pocket watch. On the white enamel face is a ring of black Roman numerals marking the hours, surrounded by another ring of Arabic numbers in intervals of five leading up to sixty. Outside this, at each quarter point, is a delicate, curlicued floral motif, and the hour hand itself ends in a finely crafted design that looks like a plant emerging from the ground. It's a beautiful piece of work, but, for me at least, the true beauty and complexity of this watch shine through when you open the case to peek inside. All bright gold, there are circles of embossed patterns similar to the clockface, forming protective shields for the delicate mechanisms. It looks more like an intricate piece of jewelry than a mechanical masterpiece.

The H4 had a mainspring and fusee to maintain the power input. The spring relayed power consistently for a day, and then needed winding with a key. One of the special features of Harrison's design was that the gears and escapement continued to receive power and tick away even while it was being wound up—a clever way to ensure that minimal time was lost during this daily procedure. He also invented a device called the spring remontoire. Since the mainspring power output can vary (even if only slightly with a fusee, Harrison was, after all, aiming for the most precise watch in the world), the remontoire is a separate spring that applies force to the escapement as uniformly as possible. The spring in his remontoire was rewound automatically every seven and a half seconds by the mainspring. If, for example, the watch ran a tiny bit faster with a fully wound spring and a tiny bit slower with a slack spring, the remontoire ensured that this cycle was repeated so quickly and frequently that, in the long term, the watch remained accurate. Another key feature of the H4 was its larger-than-usual balance wheel and hairspring, which made it less sensitive to physical disturbances, and able to beat five times a second. When the H4 was tested on a journey from England to Jamaica, it was found that, after adjustment for its consistent and predicted daily loss of 2.66

seconds, the watch only lost one minute and 54.5 seconds during the entire 147-day voyage.

Harrison's marine chronometers were extremely expensive, so it was another century before they were used commonly; before that, only wealthy shipowners or captains could use them. In time, however, chronometers led to a marked improvement in the accuracy of maps, which in turn reduced the number of shipwrecks caused by poorly calculated longitudes. Better world maps led to a thriving shipping industry in the West, which increased trade—although, as with many engineering innovations, the consequences are not necessarily all beneficial: better navigation also led to the invasion of new lands, the expansion of the British Empire, and the fostering of the Atlantic trade of enslaved Africans, the effects of which are still seen globally today.

For centuries, spring-laden clocks and watches were the best timekeepers we had. But, as we know, time doesn't stand still, and neither does engineering. In the twentieth century, oscillating quartz crystals were developed, followed by atomic clocks. In comparison to the spring-driven escapements that typically beat six times a second, atoms in atomic clocks oscillate over nine billion times a second, providing an incredibly precise and standardized time that is now relied on by the internet, the stock markets, electrical power grids, the railway, traffic lights, global navigation systems, mobile phones, radio, and more, all around the world. Atomic clocks are expected to lose one second in about 100 million years.

But don't call time on the spring just yet: they still drive the vast majority of mechanical wristwatches, and the tradition of reinventing and refining them continues. The hairspring may be about to undergo another metamorphosis. I spoke to engineer Kiran Shekar in the US, who is developing silicon hairsprings. He explained that metals are affected by temperature and magnetic fields, which cause changes in the hairsprings and, by extension, inaccuracy in watches, but if you use specially designed compounds of silicon instead of metal, these issues are drastically reduced. Although development is still in its infancy, Kiran is excited by the many possibilities that using this new material in an old technology can bring to the size and accuracy of the next generation of watches.

In a twist of irony, just as clocks were only accessible to the elite until springs brought them to the masses, today the masses (myself included) tend to wear the more economical watches driven by electricity, while watches with springs find themselves at the high (and highly expensive) end of the market. Having now scratched the surface of the history, crafting, and science that make mechanical watches possible, I'm completely bowled over by what magnificent pieces of technology they are. Kiran said to me, "I see a mechanical watch as an exceptional example of engineering that I get to carry around on my wrist twenty-four-seven. How many things can you say that about?" He's right. After all, I don't get to carry my buildings or bridges around with me. I think it's about time I got myself a mechanical watch. (I might need to save some money, though.)

Watching Dr. Struthers delicately dismantling timepieces in her workshop with a sharp-tipped tweezer and a loupe lens propped on her cheekbone made me think of the springs I've used in my work in structures, and how utterly different they are from those used by watchmakers. The springs in mechanical watches are tiny (since even a fraction of a millimeter can severely affect their performance), whereas the springs that I come across as a structural engineer can be enormous, large enough to wrap around my thigh. Me jumping energetically on top of one would be like a fly on a weighing scale. (No engineers were harmed in this thought experiment.) In watches, one role of springs is to regulate, and they expand and contract continuously. But in structures, springs are there to disrupt. They sit still, hidden and dormant, until they are needed.

Buildings and bridges are affected by vibrations—you may well have felt the rumblings caused by a large truck driving by your office building, or someone vacuuming the rugs in a neighboring apartment. Sometimes these vibrations are infrequent, if mildly annoying, occurrences. Other times, the vibrations and noise can be frequent, regular, and disruptive, and can cause us stress, or even damage a structure. This is where springs spring into action. They are used to intercept these vibrations, absorb as much of their energy as possible, and minimize their travel into the main structure to create a nice, quiet environment.

Without knowing it, I had experimented with using springiness to reduce vibrations in my teens. Living in student housing at my university meant not only being able to hear the music from the rooms around me, but also physically feeling it. Thundering bass from amplifiers shuddered through the concrete floors and columns of my brutalist block, causing my desk to tremble while my frazzled brain tried to figure out physics sums. Music—any sound, really—is mainly transmitted through air molecules jumping around and passing their energy to their neighbors; in other words, creating a wave, which arrives at our ears and makes our eardrums vibrate. Sound also makes liquids and solids vibrate—hence me holding my head in my hands at my university desk. In fact, the more tightly knit the atomic structure of a material, the more efficient the transfer of this energy or vibrations. In general, this means vibrations travel farther and more quickly through solids. But if you layer materials with different densities to disrupt the direct pathway that is transmitting the sound, the vibrations begin to lose energy and might even disappear. (So, putting a thick wedge of coasters under each leg of my desk resulted in some, albeit limited, success.) And it was this idea—this layering of softer stuff within the harder stuff to break up vibrations—that ultimately led engineers to use springs to protect structures.

Even within the realm of vibrations in structures, springs play many roles. At their most extreme, immense coiled steel springs absorb much of the destructive shaking a skyscraper experiences during an earthquake. Smaller versions stop bridges moving dangerously when buffeted by wind or driven over by trucks, and they reduce noise in buildings near rumbling railway lines. (Two of my own projects had vibration isolation bearings incorporated for different reasons. My first ever project, the Northumbria University footbridge in Newcastle, UK, needed large springs placed below the main deck to absorb energy from walking pedestrians in order to prevent it from wobbling. A London apartment block I worked on later in my career was positioned directly above the underground train system. To absorb vibrations from the trains and stop them from percolating into the apartments, we placed large rubber bearings below the main columns of the building's structure.)

At their smallest—which is still much larger than a watch spring, closer to the size of a fist—springs flex under the grinding, throbbing motion of large machines like air-conditioning units and water pumps—the blood vessels of our buildings—to make the sur-

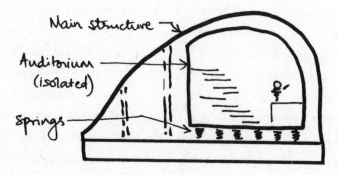

An auditorium isolated from the main structure using springs

rounding building atmosphere as peaceful as possible. It is in this seemingly small-scale role that springs have a big impact on some of the world's most famous concert halls.

The Musikkens Hus (or House of Music) in Aalborg, Denmark, looks like a large, dystopian creature. Its head, an enormous concrete cube with a transparent glass face on one side, is perched high on top of a curving concrete collar. Its body is a large U-shaped concrete box wrapped around the head, from which extend skinny circular columns that look like they will start walking at any moment. Opened in 2014, it has attracted musicians, audiences, and students from all around the world seeking to immerse themselves in its magnificent acoustics—acoustics that depend on springs.

The acoustic designer for the venue explained to me, rather poetically, that their aim for the venue was to achieve silence. Silence is one of the greatest luxuries of the modern age. All music comes from silence; there is drama in silence. It allows the audience to feel the quietest of piano notes, and the tinkle of the triangle in the back. In today's world, with endless content available online, there's something increasingly special about creating experiences that can't be had anywhere else. But this is not a simple thing to do when concert halls have all sorts of other sounds in their fabric, both from the outside—vehicles, trains, the hustle of the city—and from the inside—large air-conditioning units, pumps, and the bustle of visitors. On top of this, you need to prevent the music from one rehearsal or performance space from bleeding into and disrupting the next. In the Musikkens Hus, this is a serious challenge, since it contains a multitude of such spaces. In the head of the "creature," large orchestras play in the main concert hall, renowned for its sweeping curved balconies that contrast with its sharp-edged exterior. In the U-shaped body, there are over sixty small rehearsal rooms for teaching. And the basement contains three performance spaces—one for rock, one for jazz, and one for classical—along with the main plant room with all its noisy machinery. To create silence in such an environment, the designers of the building have included over 6,000 springs.

To prevent those listening to a Bach concert from hearing a Queen cover band next door, they created what's called a "box-in-box" system. The main structure is the first box, made from concrete. The second box, which is there to make the acoustics perfect, is built inside the main structure with an air gap in between. This sort of construction means that when a guitarist is strumming away, the sound coming from the instrument can hit any surface, but gets disrupted by the air gap because of the change of density between concrete and air. The challenge with box-in-box systems is that the air gaps have to be perforated with some form of fixing to hold the walls, ceilings, and slabs in place. If these fixings are solid, then they will transmit the vibrations, rendering the whole system pointless. This is where springs come in. The ceilings for the inner box are boards hung off coiled or curved metal springs. The walls are boards, with rectangles of flexible rubber pads (think giant erasers) joining them to the outer walls. And the floors are called floating floors. They are made from concrete and hover above the main structural slab.

The springs that hold up the floating floors at the Musikkens Hus are an ingenious piece of technology. It is not straightforward to try and form a second concrete slab above an existing one with an air gap (imagine trying to pour wet batter for a cake on top of a cooked layer with a gap in between; even with cake dowels, it's a puzzle). Instead, a thin plastic sheet was laid over the first completed slab, along with some cunning little devices called jack-up bearings.

These bearings consisted of a thick, cylindrical piece of black rubber, whose dimensions depended on making sure they would last for a long time: around, say, 150 mm in diameter and 50 mm thick, to give you an indication. At the top of the rubber was a metal screw protruding upward. A metal cup, like an upside-down funnel with a threaded spout, was rotated onto the screw. The bearings were laid out over the slab in a regular grid of around 1.3 meters. Then, wet concrete was poured on top to form the second slab. Once the second layer of concrete had hardened, a tool resembling a large Allen key was used to turn the screw in the jack-up bearings around. As it rotated, the threads, which were meshed in with the metal cup, forced the cup to lift up, dragging the concrete slab with it. Carefully and in small increments, every jack-up bearing was adjusted to exactly the height needed to create a flat floating floor with an air gap and flexible pieces of rubber that behave like springs, creating the perfect environment for immersive sound.

This might seem like a very modern piece of engineering, but in fact the jack-up bearing used at the Musikkens Hus is an evolution of a design developed in the 1960s to meet the needs of a television recording studio. At the time, CBS, a large broadcasting studio in the US, needed a floating floor that could be quickly installed in their New York City studios to improve the quality of sound during filming (some scenes they were recording involved elephants traversing the studio, and the floor needed to hold them up). On June 27, 1962, a young man sent in his solution: the sprung jack-up bearing. When I looked at the original drawing, it was so precisely drawn that I thought at first it must have been done on a computer, but the slightly slanted handwritten notes, signed with the initials "NM," showed that this wasn't the case. It was just a typically punctilious piece of work from a very talented engineer.

"NM" was Norman Mason. Born in 1925 in the Bronx, he went to a technical school in New York, but only a couple of years in, he told his parents he wanted to serve in the navy during the war. His training led him to the engine rooms of merchant ships, where he made sure that the steam- or diesel-powered engines were running

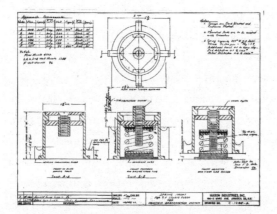

A spring mount drawn by Norm Mason.
Image courtesy of Mason Industries Incorporated

as they should, and he thrived on working on them with his hands. After he was discharged, he returned, a dashing young man with a thin mustache and a naval uniform he wore proudly, to finish his degree in mechanical engineering.

Norm, as he was known, recalls that 1948 was a terrible year for engineers, since jobs were hard to come by. He was lucky to have a lucrative but temporary job as a night engineer at the port while he looked for a more permanent position, saying that he "earned $85 a week for sleeping, while hunting for a $26 job," the starting salary for an engineer at the time. Norm's father suggested he put an ad in the "positions wanted" column instead of depending on agencies or answering advertisements for work. A stubborn young Norm disagreed with this idea, but his father posted one anyway, an act that would define Norm's future.

Just as was done before Hooke established how springs work, the use of isolation in structures was at first based on trial and error rather than science. Isolation was mainly used under machinery. As buildings grew taller and needed larger equipment to pump air and water around them, people complained about disturbance. So, cork or rubber pads were placed retrospectively below machinery to minimize noise traveling into surrounding rooms (just like the coasters underneath my desk legs). Even though the building owners in question had a problem to solve, it was tricky to convince them to spend money on this unproven technology, so a firm called Korfund (*"korfund"* being the German abbreviation for "cork foundation") found another way. Instead of simply trying to sell their products, they studied the machines that were causing the issue, provided a quote to install their solution, and guaranteed to fix the problem or not get paid.

One of the twenty responses Norm received to the ad his father posted was from the Korfund Company. Norm's stubborn youth nearly got the best of him again—he stood up and almost left his disastrous interview with them halfway through, but was convinced to stay by one of his interviewers. When Norm realized that the job entailed working on machinery, not just designing things at a desk, he accepted. As he built up his experience, he craved the independence to engineer and manufacture the best possible products, and in 1958, ten years into his career, Norm set out on his own, selling his designs from the trunk of his big Buick, with a six-page brochure held together with brass tacks. It was a modest but determined start for his business, Mason Industries, which today has a presence in close to fifty countries globally.

Norm described the decades following the Second World War as a general conspiracy to embarrass the vibration-control engineers. Postwar, people needed—and wanted—better accommodations. Entire buildings started being air conditioned. Machinery had typically been placed in the basements, but basements became cash-generating car garages, and the equipment needed to be placed elsewhere. Then, computers came, and engineers could quickly complete calculations that we struggled to perform manually, leading to lighter and more geometrically convoluted towers. Expanding populations and a trend for migration to cities led to a denser urban landscape, with the inevitable result of building closer and closer to our infrastructure. Sometimes, new trains and tunnels were even installed underneath existing buildings. Norm acknowledged that the unscientific—medieval, as he called it—method of using cork or rubber pads had served its purpose in bringing the concept of vibration isolation to the market, but that after being successful for a couple of decades, the vibration-control industry was flailing. It was time to get scientific.

For vibrations to be absorbed well, a piece of machinery needed pads or supports that were more flexible relative to the floor below it. The issue with cork and thin rubber pads was that they weren't particularly springy; in other words, they didn't compress much when squashed. This wasn't much of a problem in basements, where these thinner pads were on top of strong slabs. Even limited flexibility was enough. But as structures became taller and needed more plant to serve water and air throughout their height, plant had to be put in at higher stories. And these floors were now being made from steel, making them lighter and more flexible, and more susceptible to vibrations. This meant that the floors moved *more* than the pads did under the weight of a machine and absorbed all the vibrations.

Norm realized that you couldn't simply select a support that suited the machine; you also needed to make sure it was compatible with the structure below. By thinking of the system as having two springs—the support itself, but also the slab—he developed equations that led him to design products with "springier" springs. In the 1960s and 1970s, coiled metal springs (which are far more flexible than cork pads) became the norm for sensitive areas in buildings. By this time, Norm had finished a number of projects successfully and enhanced the theory with real data.

As Norm sought to develop the science of sound, so too did

the community at large. The intervening decades were a flurry of study to further deepen our understanding of how sound and movement interact with structure. Today, engineers designing the Musikkens Hus and other structures use a whole host of sophisticated tools to run simulations to predict how a space might sound and what interventions might be needed. By taking live measurements of vibrations from the surroundings, and anticipating the internal ones, the acoustic isolation world has become a proactive one rather than reactive, as it was in Norm's early days. So, when needed, springs in all their forms are included in the design from the start. Even with all this progress, versions of Norm's jack-up bearing from 1962 are still being used today. A perfect example of enduring design: the mark of a truly great engineer.

The use of springs to mitigate sound in structures has changed the way we design and plan our cities. Without springs, there are two possible scenarios. The city as we know it today, with its tall towers and apartment blocks close to trains, but with high noise and vibration levels. Concert halls that echo and have background rumbling. Hospitals where surgeons and researchers struggle to work. Stressed and sleep-deprived people. Or, a city of compromise, with homes away from infrastructure, leading to long commutes, and venues that have to separate plant and usable rooms, meaning expensive pumps and a lack of space, and limits on the height and depth of what we can build. But, thanks to springs, we can shape silence. We can build dense cities to reduce urban sprawl. We can build new things underground without severely impacting what's already there. Although hidden, springs have had a lasting impact, not only on a huge range of the small stuff that we use every day, in addition to our larger machines and factories, but also on the very shape of our urban landscape.

Magnet

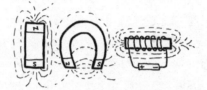

A message had arrived at the telegram office that morning. As the mailman approached the seaside apartment in Mumbai, India, that my grandfather Brij Kishore shared with my grandmother Chandrakanta and their four children, he felt his throat tighten as she pulled on his sleeve and said, *"Taar aaya hai."* In Bombay in the 1960s, the arrival of a *"taar"*—a telegram—usually meant bad news. Few homes had telephones, so far-flung family would send updates of their children, cooking, and cricket scores via the well-named snail mail. Only if a matter was more pressing and urgent would they send the news via telegram.

Babuji, as we all called him, tore open the envelope and took out a sheet of pale blue paper. On it was glued a strip of white paper that contained three words: ANXIOUS TO RETURN. He looked at his wife, rolled his eyes, and reassured her that there was nothing to worry about.

After graduating college, Babuji's son Shekhar had traveled to Italy to look for work. Evidently, he didn't like it there and wanted to return, but Babuji was determined that Shekhar should give it a shot. So, he put on his chappals and walked down to the post office to send a telegram saying so, using as few words as possible, because telegrams weren't cheap and were charged by length. Over the next few weeks, many more telegrams arrived from Italy, begging for a ticket back to Bombay. After ignoring many of them, Babuji finally relented. His son, my uncle, returned to Bombay, where he lived out his days.

Less than sixty years later, each week of pandemic lockdown was punctuated by the demanding squeaks of my toddler: "I want talk Nani right now!" The child of a pandemic, there were eighteen months of her life when she couldn't see her grandparents in person, so her demands to speak to her grandma were swiftly obeyed. With the swish of a finger on a touchscreen, a call flew through the air to the other side of the planet, to which my mom responded. She saw my daughter crawl for the first time, and speak her early words, in color, live, on the screen of a smartphone. When I stop to think about the ease with which we were able to stay in touch through those tough times, I find myself not only in awe of how far we have come but also immensely grateful.

We have been through a radical shift in technology across just three generations of my family, and each step of the way has changed our lives dramatically, just as they did for society as a whole: allowing us to communicate with our loved ones, creating the world of instant news, changing the way we work, and altering the way we entertain and are entertained. But while a video call may seem a far cry from the telegram, all these forms of modern communication are based on the science of signals being sent from one distant point to another, almost instantaneously. And our ability to do that centers around magnets.

I find magnets magical. The magnetic fields that radiate from them are invisible, but they can be substantial, far-reaching, and influential across large distances. The science is complex and wasn't understood for thousands of years—indeed, many physicists will tell you that magnetism, and especially electromagnetism, still isn't fully understood. But once we had at least some understanding, we were able to create practical mechanisms. Humans harnessed the magic of magnets to create machines that could interact and exert forces on other machines, farther away than had ever been thought possible.

Unlike the inventions we've looked at so far, magnets—or objects that exert magnetic forces—exist in the very essence of our universe. You and I are magnets (very, very weak ones—don't worry, there's no danger of us suddenly becoming attached to our refrigerators). Atoms, the minuscule building blocks of matter, are magnetic. The planet on which we live is a giant magnet. Magnets, unlike wheels and nails and springs, were discovered rather than invented by humans. Despite this, they nonetheless deserve their place in this book, because it was humans who figured out how to make them more useful than they were as supplied by Mother Nature. The magnets we found naturally in our surroundings a few thousand years ago were weak and hard to come by. They were formed of magnetite, which came to be known as lodestone, a natural mineral found in the earth that is a mix of iron and oxygen, plus other impurities. It's a magnetic material, but only a small proportion of the magnetite that exists in nature is magnetic, because it needs both a specific combination of impurities inside it, and to have been exposed to specific conditions of heat and magnetic fields outside it.

The earliest references to this natural magnet date back to ancient Greece in the sixth century BCE. Around two hundred years later, the Chinese documented the phenomenon of a natural stone attracting iron, and in another four hundred years, they began using this material for geomancy (a form of divination). It took another thousand years, advancing into the Middle Ages, before it was used for navigation in the form of a compass. Navigators in the Song Dynasty in China shaped lodestone to look like a fish, and let it float freely in water, so it pointed south. This knowledge spread to Europe and the Middle East soon after. Even then, with over a thousand years of knowing about natural magnets, we couldn't replicate them, and their use was restricted to navigation.

Magnets themselves come in two distinct forms: permanent magnets and electromagnets. Permanent magnets are the horseshoe- and bar-shaped magnets we saw in school science demonstrations and those that decorate our refrigerators. They have two poles, north and south: bringing together the south poles or the north poles of two magnets creates a pushing or repulsion force, but bring a north and south pole together and the magnets will cling to each other.

It took millennia to come to grips with how magnetism works, because this requires an advanced understanding of atomic physics and material science. To become a magnet, a material requires many particles, at many different scales, behaving in a very particular way. Let's start with the electrons that orbit the nucleus of an atom. Just as electrons have a negative electric charge, they also have what physicists call spin, which defines its magnetic characteristics. By "pointing" in different directions, the spin cancels out the magnetic forces of electrons entirely in some atoms, leaving them nonmagnetic. But in others, while some of the electrons are arranged so their spin cancels out, not all are, so there is a net magnetic force left over, creating a magnetic atom.

Then, if we zoom out from the electron scale to the atomic scale, the atoms in an element are naturally arranged at random, which means that the magnetic forces of the individual atoms cancel each other out. In some materials, however, little pockets of atoms— called domains—have atoms all arranged in the same direction, giving the domain a net magnetism. However, they are not yet magnets, because the domains themselves are usually arranged at random.

To make a material produce a net magnetism, then, the atoms in the majority of the domains need to be forced into magnetic alignment by a strong external magnetic field, or by large amounts of heat applied at particular temperatures in particular sequences. Once the *domains* point in the same direction, you have a magnet.

Even today, there is a debate as to how magnetite becomes magnetized in the first place, so artificially replicating this has been a challenge. Certain materials like iron, cobalt, and nickel have electrons favorably arranged to make their atoms magnetic, which in turn sit in well-defined domains. Our ancestors tinkered with mixes of such metals, heating and cooling them in various combinations to try to figure out the best recipe for forming permanent magnets. They succeeded, to a degree, making somewhat weak magnets that didn't hold their force for long.

The development of permanent magnets in a scientific way started in the seventeenth century, when Dr. William Gilbert published *De Magnete*, which outlined his experimentation with magnetic materials. In the eighteenth and nineteenth centuries, we developed more sophisticated methods for making iron and steel, and observed that certain combinations made much stronger or longer-lasting magnets—and sometimes even both. But we still didn't really understand why. The nineteenth century also saw the advent of understanding electromagnetism, which we'll come back to, but it took until the twentieth century and the conception of quantum physics before we were able to define and understand atoms and electrons well enough to create strong and long-lasting permanent magnets ourselves.

This led to the use of three types of materials to make permanent magnets: metals, ceramics, and rare-earth minerals. The first major improvement was the development of a metal mix of aluminum-nickel-cobalt, used to make "alnico" magnets, but these were complicated and expensive to make. Then in the 1940s, ceramic magnets were created from pressing together tiny balls of barium or strontium with iron. These were much cheaper, and today account for the vast majority of permanent magnets produced by weight. The third family of materials are the rare-earth magnets, based on elements like samarium, cerium, yttrium, praseodymium, and others.

Within the space of the last century, these three types of permanent magnets have been refined to produce produced magnetic fields 200 times stronger than before. And this improved efficiency led to permanent magnets playing an important role in much of our modern lives: a car, for example, can have thirty separate applications for magnets, using over 100 individual magnets. Thermostats, door latches, speakers, motors, brakes, generators, body scanners, electric circuitry and components—take any of these apart and you'll find permanent magnets.

But as we saw, the stories of permanent magnets and electromagnets intertwine, and since the discovery of electromagnets around 200 years ago, each has swung in and out of favor as humanity learned more about how they worked and what they could be used for. The prevalence of permanent magnets in the past few decades is due not just to their increasing strength and compactness but also to the fact that, unlike electromagnets, they never need a source of power. But from the nineteenth century onward, and even today in situations where immense fields are needed, electromagnets dominated. We can control their strength, switching off or cranking up the magnetic field of an electromagnet when it suits.

The reason electromagnets took so long to make an appearance in the field is because we needed an understanding of the science of materials, electricity, and light—and the mysterious force of electromagnetism. It's only when we were able to move electrons in materials that we understood how to create and change this force and apply it to our technology.

Like gravity, electromagnetism is one of the fundamental forces in nature. It is the physical interaction that happens between particles, like electrons, that have an electric charge. In the late eighteenth and early nineteenth centuries, André-Marie Ampère, Michael Faraday, and other scientists published numerous theories about electric and magnetic fields, which were eventually brought together and summarized by the mathematician James Clerk Maxwell in what are now known as "Maxwell's equations." These gave us crucial information that led to the invention of electric motors, and these equations are also the basis of our power grids, radios, telephones, printers, air conditioners, hard drives, and data-storage devices; they are even used in the creation of powerful microscopes.

The key principle that led to such technological advancement was the realization that moving charges create magnetic fields. Without getting too deep into the complex science, this means that if an electric current is flowing through a coil of wire, it behaves like a magnet. If you change the strength of the current, you change the strength of the magnet. And the converse is also true: applying a variable magnetic field near a wire will create an electric current in the wire. Following on from this science, experiments proved that when a charge, like an electron, moves within a magnetic field (either freely or inside a wire), it feels a pushing force.

Studying the electromagnetic force led us to define the phenomenon of electromagnetic waves. Think of these as waves of force that flow because of the interaction between electric and magnetic fields. Our understanding of light increased manifold when we were able to quantify it as an electromagnetic wave (more on this in the next chapter). And, in addition to visible light, we saw that a whole spectrum of electromagnetic waves— from radio waves (with the longest wavelength) to gamma rays (with the shortest)—exists, and that these waves can be used in different ways. It is electromagnetism and electromagnetic waves that form the basis of our long-range communication technology: the technology used by countless people around the world to share news with their loved ones. People like my uncle, the prolific sender of telegrams.

For tens of thousands of years, humans couldn't send messages quickly over long distances. Communication meant physically transporting themselves or their letters, using runners or people on horseback to make arduous journeys, until the wheel came along and sped things up—slightly. In time, ancient Chinese and ancient Egyptian civilizations found ways of sending signals relatively quickly over fairly long distances by putting up pillars at regular intervals to send out smoke signals, or stationing people to beat large drums to send coded messages when danger was imminent. These methods of conveying messages evolved into the semaphore system, which relied on a series of tall towers topped with two telescopes, each one pointing toward the nearest tower on either side. Operators in one tower would communicate using a language composed of shapes made by flags held at different positions or angles, which were observed via the telescopes on the adjacent tower, and then passed on to the next. Though useful in an emergency, these systems were very limited; they weren't particularly far-reaching, not only in terms of distance and the portion of the population that could send messages, but also in terms of what messages they could actually communicate. The possibility of people communicating across lands in a matter of hours or minutes only became a reality in the last 200 years, when electricity and magnetism had become better understood, leading to the invention of the telegraph.

The ability to send information telegraphically changed the way we live. Trade became less risky and variable, as merchants could keep track of prices hourly. The railways could send alerts of the progress of steam engines to keep the network running safely. And news reporting in the way we know it today became possible, because people no longer had to wait days or weeks to hear the latest developments; now, journalists could relay breaking news across distances. It brought people together, allowing families to keep in touch and stay in contact with friends who were far away. But easier communication meant easier subjugation; colonists could keep a tighter grip on their colonies.

The telegraph system that delivered my uncle's messages had its origins in the British Empire. The geography of India posed a big challenge for any method of long-distance communication. Unlike in Europe, the terrain varied from mountains and jungles to swamps and unbridged rivers, with large birds and monkeys and thunderstorms all creating difficulties for European engineers. Initially, a visual system based on the semaphore design was completed in 1818 in the eastern part of the country to facilitate quick communication between Calcutta and Chunar, a distance of nearly 700 km.

Then, in 1839, a twenty-nine-year-old British doctor, chemist, and inventor named William Brooke O'Shaughnessy constructed an experimental electric telegraph line in Calcutta. An iron wire was strung back and forth across a series of vertical bamboo poles, with the transmitters and receivers of the line brought together at the same end for ease of the experiment. Thanks to experimentation in electricity and the invention of the battery in preceding decades, running current through the cable wasn't a concern. The concern was how this current would be translated into language.

O'Shaughnessy worked independently of the British (William Cooke and Charles Wheatstone) and American engineers (including Samuel Morse) who were designing their own systems in the 1830s. One of his first designs called for messages to be sent based on two clocklike devices. The sender and receiver would both have the same "clock," where the dial had letters rather than numbers. (He considered using chronometers, but decided they were too expensive to buy, and then mess with, to suit his experiments.)

His design involved the sender transmitting a pulse of current when the hand of the clock was pointing toward a particular letter, which would physically shock the hands of the receiving operator, who would look at the clock and note down the letter highlighted in that instant. O'Shaughnessy described the sensations of these shocks in his accounts of his experiments in 1839—ranging from those that are "utterly intolerable" and "strong, but not disagreeable sensations," to the feeling of a "blunt saw drawn lightly across the palms." O'Shaughnessy tried to justify his system by saying that the "eye and ear are liable to distraction," but "the attentive touch knows no interruption."

Luckily for his operators, the science of electromagnetism was becoming increasingly understood. The Danish physicist Hans Christian Ørsted had theorized in 1820 that when a compass was placed near a wire with electrical current, the magnetic needle of the compass spun a little and ended up pointing in a different direction. O'Shaughnessy used this principle and tested different designs that contained electromagnets.

He recorded how quickly signals traveled using different materials and sizes of wire, and tried out different arrangements of needles close to coiled wire. He settled on a system where a current ran through the coil, produced a magnetic field, and caused a needle to swing either right or left, depending on the direction of the current transmitted. He called these deflections "right-hand beats" or "left-hand beats" and defined each letter of the alphabet with a series of beats. So, A was one right-hand beat, and B was two right-hand beats, going all the way through to Z, which was four left-hand beats and one right-hand beat.

With this proof of concept—a telegraph where magnets made messages transmitted by electricity decipherable—O'Shaughnessy shared his ambition of a cross-India system. Governor-General Dalhousie helped him circumvent the cumbersome bureaucracy of the East India Company to get a route approved in the early 1850s. In 1852, the first section of line was completed between Calcutta, Diamond Harbour, and Kedgeree on the River Hooghly, under the management of O'Shaughnessy's protégé, Seebchunder Nandy (the first Indian to make an appearance in official Telegraph Department records). Nandy, who sent the first signal from the Diamond Harbour end, which was received by O'Shaughnessy in the presence of Lord Dalhousie, was then appointed inspector of the line.

Between 1853 and 1856, 6,500 km of line was constructed. The route was operated via sixty stations, starting in Calcutta in the east, running northwest to Agra, then south to Bombay along the west coast, and across to Madras on the east coast, with a branch through Delhi and Punjab in the north. Since the line was laid out based on colonial political and military tactics, the territory along the east coast between Calcutta and Madras was left out because it was not considered important. For similar reasons, construction was planned in two stages: the first was a temporary line of bamboo poles to quickly establish a communication system for the Imperial forces, who were on alert for local uprisings. Then, permanent poles were installed, which were secured by shaping one end like a large screw that was twisted into the ground. Messages going through the system were dominated by government and military correspondence, but from 1855, the public were allowed for the first time to use the telegraphs, at the rate of one rupee (around 2,500 rupees, or about 30 US dollars today) for a message of up to sixteen words over a distance of 400 miles.

The system had a mixed experience during the First War of Independence, when Indian sepoys in the British Army rebelled against the British in 1857. The revolutionaries destroyed nearly 1,500 km of the line to sabotage British attempts to alert bases around the country and took advantage of the areas not covered by the telegraph. However, the British managed to get a warning of the emergency to stations in the north and south just in time, allowing them to limit the rebellion to a large extent, rendering it unsuccessful. The network was then expanded and upgraded to cover nearly 18,000 km, routed to suit the colonists' needs. By the end of the 1860s, the system had reached London, helping the Imperial government establish tight control over its colony.

By 1939, still under British rule, India had 100,000 miles of line carrying 17 million telegraphic messages a year. But the system reached its peak in the years after Indian Independence in 1947: in 1985 (two years after I was born), 60 million telegrams were sent and received from 45,000 offices. But, of course, technology never stands still, and the development of mobile phone technology and the internet eventually led to the closure of the service. The last ever telegram in India was sent on July 14, 2013.

I know that at least three generations of my family used telegrams. Even though the electronic telegraph is almost obsolete today, it marked a turning point in the story of rapid long-distance communication, which all began when someone dreamed up the possibility that a magnet could be used to convert electrical pulses into written language. This story of the telegraph also reminds us of the ways in which wondrous scientific advancements can be exploited by those with power to hold on to their power and to oppress those who don't have any.

Lata, my aunt, and Vijay had met at college, where they were part of the same band; Lata was the lead singer and Vijay played the tabla. Practicing music together helped relieve some of the stress of their medical degrees and nurtured a romance. They got married soon after graduating, and in 1970 decided to emigrate from India to the United States to establish their medical careers.

Vijay spent frequent sleepless nights on call in the emergency room while training as a surgeon, and Lata worked long days in the hospital's gastroenterology department. Life was difficult and lonely for the young couple without their family and childhood friends—they were the first in their homes to graduate, and the first to set up life in an unknown land—and they craved contact with their loved ones. But that contact wasn't easy. Once a month, Lata would call up the telephone exchange in New Haven, Connecticut, and book a trunk call to Bombay for $5. She gave the operator her family's home phone number. Twenty-four hours later, the New Haven exchange got back in touch with Lata with a date and time for the call.

Lata only paid for three minutes of talk time. Having already spent $5 simply to book the call, she then had to pay another $15 for three minutes, which added up to a lot of money (around $140 today). The first few calls she made through the crackling wires of the telephone exchanges only really allowed her to faintly hear the voices of her parents, Babuji and Maa, and her siblings, including my father, as they all shouted "Hello" loudly into the receiver. They were barely able to hear each other before being abruptly cut off when their time was up. Eventually, they made a pact to ban saying hello and asking each other how they were, which made room for a rapid exchange of key updates. The more detailed conversations between this generation of my family still relied on the long, lyrical letters that my grandfather and aunt exchanged each fortnight.

Letters worked because we had the means to physically transport pieces of paper across continents. Transmitting sound over
such distances wasn't as straightforward, because sound is a vibration. When we speak, air passes through muscles in our voice boxes
and makes them flutter, creating a wave of vibration that travels
out into the air and into our eardrums. The eardrums then vibrate
in turn, and thanks to the brain interpreting these signals, we
hear. But our voices can only travel so far, because as vibrations
travel, they lose energy (just as ripples from a pebble tossed into a
pond eventually disappear).

To increase this distance, you could make a crude telephone
with two empty tin cans joined with a piece of string. If two people take a can each and move as far away from each other as the
string allows, they can take turns hearing each other's voices rat-
tling through the cans. The person speaking causes vibrations in
their can. These vibrations travel through the taut string to the
second can, which begins to vibrate and emit sound. The string is
pretty efficient at letting vibrations travel without losing too much
of their energy, but these machines have their limits in distance—
and in practicality.

Although the telegraph was breaking these limits, it was done in
a way that needed some sort of code to be interpreted: you couldn't
speak into a telegraph. In order to take communication to the next
level, we needed to merge tin-can-phone and telegraph technology.

The telegraph—which uses a magnet to convert pulses of current into the movement of a needle—inspired engineers to think
bigger. In the 1870s, we theorized that if an electric current could
create movement in a magnet, and if, unlike the telegraph, this
movement happened hundreds or thousands of times a second,
then a vibration would be created. This movement, or vibration,
could produce sound.

Early telephones came in different designs, but they had one thing in common. They used magnets, and coils of wire together, to create an interaction. Electric currents and magnetic fields are so intertwined that a change in a current produces a change in a magnetic field, and a change in a magnetic field causes a change in a current. Alexander Graham Bell, an American inventor who was awarded the patent for the first telephone (despite someone else filing paperwork just hours later), experimented with different arrangements of magnets and coils to create his instruments.

Bell's mother and wife were both deaf, and his life's ambition was to teach deaf people how to speak. In her book *The Invention of Miracles*, Emily Booth explains that, rather than a telephone, Bell was originally trying to create a machine that would translate the vibrations of speech into something visual that deaf people could see. However, his legacy among the deaf community is checkered because he fought really hard for education that eliminated sign language from the classroom, and, in the same year as the word "eugenics" was coined, he published a memoir that advocated against deaf people marrying each other. He was very concerned that deaf people were having more deaf children, which might lead to a whole population of the human race that was deaf. Booth says that it's a complicated story, because Bell believed he was doing the right thing, but his dual mission to teach the deaf to speak and to eradicate sign language still adversely impacts education for the deaf today.

Ironically, Bell ended up creating a device that the deaf couldn't use. One of his early designs had a long, permanent bar magnet laid on its side, supported by two wooden legs on top of a wooden block. Attached to the end of the magnet was a coil of fine wire. Just beyond the same end, a thin iron disc oriented vertically was fixed onto a separate wooden leg, leaving a small gap between the disc and the coil. Since the disc was made from a magnetic material, a field was formed between it and the bar magnet. A funnel

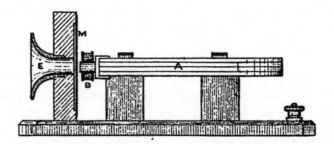

One version of Alexander Graham Bell's telephone showing: A: perma-
nent bar magnet; B: small coil; M: a thin iron disk; E: mouth and
earpiece. From A Manual of Telephony *by William Henry Preece FRS*
and Arthur J Stubbs, published by Whittaker & Co. 1893

near the iron disc (or diaphragm) transmitted sound waves back
and forth. An identical second instrument was connected to the
first by a wire.

The way Bell's telephone worked was that you spoke in front
of the funnel (in this case performing the role of a mouthpiece),
which sent vibrations to the diaphragm, which mimicked the pitch
and loudness of your voice. The diaphragm vibrated, moving back
and forth, and the movement of the diaphragm relative to the bar
magnet made rapid changes to the magnetic field between them.
The rapidly varying magnetic field induced a rapidly varying electric current in the coil, which ran through the wire to the second
instrument. Here, the opposite happened. The same variable current as the first instrument was reproduced in the coil, which in
turn generated a varying magnetic field, which pushed the diaphragm back and forth, making it vibrate exactly as the receiving
end had—although less strongly, as the vibration had naturally
lost some of its energy. The second person placed their ear on the
funnel—now an earpiece—to receive these weakened vibrations
from the diaphragm in their ear. They could hence (just about)
hear what the other person was saying.

On February 12, 1877, in the Lyceum Hall in Salem, Massachusetts, Bell used a modified version of this instrument to call his assistant, Thomas Watson, sixteen miles away in Boston. Bell alerted Watson that he was ready by knocking on a small thumper that hit the instrument's diaphragm (an innovation thought up by Watson), to which his assistant responded in kind. Then, Bell leaned close to the box and asked, "Mr. Watson, can you hear me?"

For a moment, only crackling came through the line, until the audience heard Watson respond, "Yes sir, I hear you." Another crackly pause, and then Watson offered to sing a song for the audience in Salem.

After the demonstrations were finished, a reporter at the Salem end used the telephone to dictate an account of that evening to a colleague in Boston. So, the story of the first ever long-distance call was relayed using a telephone and was published the next morning under the headline "The First Newspaper Despatch Sent by a Human Voice Over the Wires."

This was a momentous leap in our ability to communicate, even if the telephone still had some idiosyncrasies. The funnel on this version of Bell's instrument was both the transmitter and the receiver, so after you spoke into it, you then had to put your ear against it to hear what the other person was saying, which was not convenient. Transforming sound into current in the receiver posed a challenge, with solutions needing a huge battery to power the system. This issue was only resolved about fifty years later in 1926, when James West and Gerhard Sessler, based in Bell Labs in the US, invented the so-called electret microphone. (Luckily, West doggedly pursued his dream of becoming a scientist despite segregation being rife, with very limited opportunities for Black men in the field.)

In addition to the issues specific to the instrument itself, there were also challenges in connecting different instruments. Telephones operated in pairs—for example, at your home and your office—and could only call each other. If you wanted to connect your home telephone to, say, three other points, you'd need to have three wires, one for each location. This might not sound unreasonable, but if everyone wanted three connections, let alone more, you can imagine the wire spaghetti that would lead to. To solve this, a different system was needed.

New Haven, from where my aunt made those calls to Bombay, was the first place in the world to have a commercial telephone exchange that acted as a central meeting point of telephone cables. (She would have connected to an updated version; the original site was pulled down in the 1970s.) At the original New Haven exchange, twenty-one subscribers had cables traveling from their businesses and homes to the exchange, where a switchboard— apparently assembled from carriage bolts, handles from teapot lids, and wire—allowed them to connect to each other. For $1.50 per month, a caller could wind a crank on their phone (which rotated a magnet relative to a coil of wire, hence creating a current) to let the exchange know they wanted to place a call. They then told the operator the number, and the operator would physically plug in jacks (like those at the ends of our headphones) to connect a cable between the right two lines to complete a circuit. Once the caller had finished their conversation, they rang the bell with the crank to let the operator know it was over. My aunt had to get several different exchanges to connect up in order to get her call in to Bombay; there was no one switchboard that housed both her line and her parents' line in India. She connected into New Haven and my grandparents connected into Bombay, but other exchanges in New York and London were needed to complete the chain. That's why it took twenty-four hours to schedule a call: the call could only be placed when the entire series of wires, or trunks as they were sometimes known, were available.

When there were relatively few people with telephones, it was practical enough to employ operators (usually women, who became known as "hello girls") to manually plug leads in and out of switchboards. However, as the numbers of people using these devices exploded, these women became overwhelmed—despite their quick hands and good interpersonal skills—which often led to people waiting on the line before a slot became available.

The story of the automation of the exchange—an invention born from sheer spite—starts at an undertaker's office. It's said that in the 1880s, Almon Strowger was the only undertaker in El Dorado, Kansas. But another soon arrived, and Strowger found his business dropped dramatically. It turned out that the wife of the new arrival worked at the exchange, and when people called asking for Strowger, she routed them to her husband's business instead. Frustrated, he wondered how to create a "girl-less, cuss-less, out-of-order-less, and wait-less" exchange.

In 1892, Strowger patented his automatic exchange, where he replaced the potentially dishonest human operator with a magnet. At the same time, he created dial telephones, those lovely vintage phones where you see a dial with holes above each number. If you wanted to call up phone number thirty-eight, you stuck your finger in the hole of your telephone above the three and rotated the dial around. As the dial rotated back into its original position (thank you, spring), it sent three electrical pulses down the line. Dialing eight sent eight pulses.

Strowger's automatic exchange was in the shape of a cylinder that had ten rows of metallic rings cased within a nonmagnetic material. Each ring had ten bumps with wires attached, and each of these bumps represented someone's telephone line. In the center

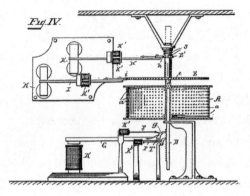

Image from Strowger's patent showing electromagnets at K and K'

of the cylinder was a rod that had teeth at its base, gears at its tip, and an arm. A lever, which was connected to a pair of electromagnets, held the rod in place vertically at the bottom by biting into those teeth, while another two levers (also connected to a pair of magnets each) meshed into the gear wheels at the top.

When the first three pulses arrived from your telephone, the electromagnet at the base turned these pulses into vibrations and caused the lever to push the rod up three steps, so its arm came into contact with the third ring. The second eight pulses sent signals to the top set of magnets, which forced the cylinder to rotate so its arm connected to the eighth bump of the ring. This completed the connection, and you could now have a conversation—with the right undertaker.

Telephone exchanges that required operators created one of the early and large female workforces, and the automatic exchange left them looking for other options or returning to managing their homes full-time. Ironically, Strowger also changed professions because of his work: his efforts to save his undertaking business resulted in him leaving it to mass-produce his new invention. With his family and friends, he created the Strowger Automatic Telephone Exchange Company, which installed the first working system into La Porte, Indiana, in 1892.

Strowger's design was expanded by placing a series of the cylinders one after another to increase the number of possible connections. Still, the system required moving parts and lots of space. In the 1940s, as a new world of electronic devices opened up thanks to the transistor, Strowger's exchange was replaced by a digital system that paved the way for our mobile phones. Lata and Vijay use this new technology now, but in their early days of marriage, in order to hear their family's voices, they relied on magnets converting sound to electricity and back again.

Lynette had lived in many places. After being born in London, she moved to Giridh in Jharkhand, India, and then to Bombay as a teenager to go to college. There, she was introduced to Hem, an engineer who worked in upstate New York, where she moved in 1978 after they got married.

An avid consumer of popular culture, she had been disappointed to find that her student hostel in Bombay had only a black-and-white television. On arrival in the US, she bought a color TV for their home, and watched many, many episodes of the game show *Wheel of Fortune*, introduced by a male voice-over excitedly listing what the contestants might win, followed by the audience shouting out the name of the program in unison.

That introduction, as well as the jaunty piano music that opens the series *Murder, She Wrote*, brought back memories from my early childhood of standing around the edge of the living room just before bedtime. Lynette, my mom, let me listen to the introductions before I was shooed off to bed. During the day, I watched videos of

Madonna performing her eighties hits, and Shammi Kapoor pranc-
ing in gardens while serenading his latest love interest.

I remember that television we had. It was about as wide as my
shoulders. When switched on, a horizontal line flashed across its
center; it took a second to warm up before casting an image across
the screen. If I stood really close to it (which I wasn't supposed to
do), I could see lots of red, green, and blue dots. It had a curved
glass screen, and the back of the set was bulbous, wrapped in a
casing of black plastic with long, thin slots. I now know that behind
this black plastic was a powerful electromagnet.

Globally, television has completely changed our daily lives. Its
development involved many technologies developed by many people from many countries around the world. The earliest television
sets couldn't have existed without this complex web of innovation,
but I'll stick with the story of one particular pioneer. He isn't well
known because he didn't patent his earliest designs; he worked
largely independently from the inventors from the West, and much
of his work and records were destroyed in the Second World War.
In his native Japan, however, he's rightly considered the Father
of Television.

Takayanagi Kenjiro was born in Hamamatsu City, Japan, in
1899. By the time he died, ninety-one years later, he had received
the highest civilian honor from the emperor, a huge achievement
for someone who almost missed out on further education.

Takayanagi's father was a failed businessman and couldn't
afford to send his son to middle school, so the future pioneer
resigned himself to a life of toil. But when he patched up the shoji
windows of the family home with some beautiful calligraphy, it
caught the eye of a school principal. Thanks to this principal, and
a childless aunt who decided to look after her nephew, he was able
to go to school, and then a technical college, before becoming an

instructor at Hamamatsu Technical High School in 1924, where he trained students to become electricians and technicians.

Takayanagi was a dreamer. He was obsessed with radio, and was known for standing at the doorways of hotels to stop foreign visitors and ask them about the latest developments. He wrote in 1924 of a vision where his family gathered in front of a single machine, on which, after adjusting a knob, a gorgeous dance from the Imperial Theatre appeared on a screen before their eyes. He wanted to create what he called a "wireless with vision."

Takayanagi managed to convince his school principal to part with a small research allowance, apparently using his wife's dowry when this ran out. Working in almost complete isolation, he succeeded in transmitting an image of the Katakana syllable for "i" from a camera in one room to a screen in another on December 25, 1926—the first electronic television in the world that used a cathode-ray tube. He didn't apply for a patent. (This was about two weeks before Philo Farnsworth, often considered the inventor of the television, filed his patent for a television system in San Francisco, California.)

The main components of the early televisions were a transmitter and receiver. The role of the transmitter was like a video camera: to capture a series of images and convert the images into a signal that could be sent onward. The receiver then received this signal and converted it back into a series of quickly changing images, to create a moving film. Although Takayanagi's ambition was to have electronic devices for both components, he wasn't able to achieve this in his 1926 television. The receiver was electronic, but the transmitter was mechanical, and this limited the quality of the image produced. Over time, he developed these key components to create the best television images of the era, by incorporating magnets into both.

To see why the quality of a mechanical transmitter is limited leads me to memories of playing with sparklers on Diwali. One of my favorite things to do was to move my arm really quickly to draw circles in the air. Although we know there is only a spot of light, the reason we see a circle is because our eyes and brains aren't good at processing fast-moving dots of light, so instead we see a blur. Televisions were designed around this principle, but instead of moving light quickly in a circle, engineers used a pattern called a raster.

This pattern is made when a spot of light starts, say, at the top-right corner of a rectangle, swiftly moves left in a straight line across the shape, and then, at the end of the rectangle, jumps back to the right end, but this time, one line down. (The pre-1990s children among us will recall early printers did a similar thing.) If this is repeated rapidly enough until the light covers the full height of the rectangle and then starts again from the top right, you can create the illusion of a rectangle full of light. Then, if you vary the intensity of the light as it flits back and forth across the screen, you can create patches of light and dark. Be really clever about choreography and these patches can coalesce to form an image, or a series of images slightly different from one split second to the next, and then you have yourself a moving picture.

The mechanical device that Takayanagi used in his first television was the Nipkow disc. The disc created a raster pattern thanks to a series of small holes drilled in a spiral around its circumference. He placed the disc vertically, with a transmitter that shone a bright and varying light behind it. As the disc spun, pulses of light popped through the holes that were rapidly moving in front of it. When light shone through the holes in the disc closest to its edge, they "drew" lines across the top of a screen, gradually moving toward the bottom of the screen as the holes became closer to the disc's center. And so, he created a raster pattern with forty lines. By 1927, he had improved the resolution to 100 lines, which was unrivaled until 1931.

Televisions like the one I watched in the 1980s typically had 480 lines, which were refreshed thirty to sixty times a second. The difference between Takayanagi's first machines and my childhood set was the magnet. To create a sharp picture that didn't flicker required a much faster and denser raster pattern; in other words, many more lines of light achieved at a very high frequency. But a rotating disc can only spin so many times a second, and can only be so large before it becomes unwieldy and difficult to maintain, so there's a limit to its quality.

In 1929, Takayanagi made extensive improvements to a device called the Braun tube, or cathode-ray tube. He changed the shape of the glass tube from one of uniform width to being funnel-shaped, using the wide end as a screen. This was coated with a chemical layer that lit up when the electrons—which, like light, travel in straight lines if there is no force acting on them—hit it. The narrow end had an electron gun—an arrangement of wires that emits a stream of electrons.

With this arrangement, you could see only a dot of light in the

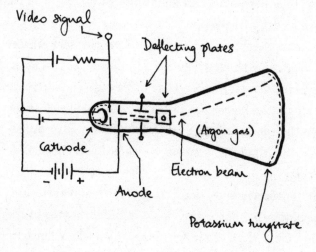

The Braun tube, adapted by Takayanagi, and used to create his television

center of the screen, and the brightness depended on the energy the electrons had, and whether they could travel cleanly through the tube. Takayanagi succeeded in increasing tenfold the energy of the electrons hitting the screen. He removed as much air as he could from the tube to stop other air or gas particles from interfering with the beam. And, importantly, he added electromagnets arranged in such a way that the magnetic field exerted forces on the electron beam and could move it rapidly in a raster pattern. This time, he wasn't restrained by the speed at which a disc could rotate—the magnetic field could be varied extraordinarily fast simply by varying an electric current through a coil. And so, he was able to produce a bright and distinct image, completing his all-electric television system in 1936. In May 1939, he achieved 441 scanning lines with twenty-five frames per second, and set up the first Japanese television broadcasting system.

The color televisions with cathode-ray tubes, like mine in the eighties, had three separate electron beams that produced three different colors when they hit the fluorescent screen: red, green, and blue, which is what I could see when standing too close to my set. Together, and at an equal intensity, the three beams created white dots. Increasing or decreasing the intensity of the different beams created different colors on the screen, like mixing paint colors on a palette. The main limitation of this form of television was the screen size, because of weight. The larger the screen you wanted, the bigger the cathode-ray tube needed to be, and the thicker the glass, in order to resist the pressure of the vacuum inside it. So, the large screens we watch today needed a different technology.

Our flatscreen televisions don't need magnets to directly create their images, but magnets were indispensable in their development. LED (light-emitting diode) televisions have thousands of tiny light bulbs that use a small amount of energy and work because of minute jumps that electrons make within an atom. Red and green LEDs had been around from the 1960s, but to create a white pixel, blue would also be needed. Engineers struggled to coax electrons to jump the even tinier amount that would result in the emission of blue light. In the 1990s, a team of physicists from Japan were able to create crystals of special materials and invent this invaluable technology.

But manufacturing blue LEDs in scale remained a challenge, since the material was very prone to defects. You could only see these defects with a state-of-the-art electron microscope that, with the use of magnets, went down to 0.1 nanometers (ten billionths of a meter). This work enabled a new generation of bright, energy-efficient white lamps and color screens. LED lamps are expected to have world-changing effects, with the potential to help more than 1.5 billion people who do not have access to electricity grids, because they are efficient enough to run on cheap, local solar power.

I signed up for my first email account around 1998—it was roma_millenium@hotmail.com—and, yes, I misspelled it. I was fifteen. I would plug a phone cable into the central processing unit to connect to the internet using a dial-up connection, looking at web pages on the display screen of a cathode-ray tube. It was the early days of home internet, and I remember once waiting an epic twenty-seven minutes for my emails to load (they were all spam).

The world of rapid global communication in which we now live relies on (in addition to the technology I've already described): large data-storage units, optical cables that transmit data at the speed of light, wireless signals that are bounced around from the surface of the planet to satellites above it, and more.

Wireless technology depends on radio waves, which are one type of electromagnetic wave. We can't see these with our eyes, but they are all around us, sending packets of data to our phones, Wi-Fi, and satellites. One of the earliest pioneers of this technology was the Indian scientist Jagadish Chandra Bose.

Bose didn't confine himself to one area of research or even one discipline. In addition to writing science fiction and studying botany (he invented the crescograph for measuring plant growth), he investigated radio microwave optics. As early as 1895, he demonstrated that electromagnetic waves could travel far and effectively. He set up a public lecture, attended by the lieutenant governor of the state, where he switched on a transmitter, which generated the waves. The waves went through the lecture room itself, through an intervening room and passage, to a third room that was 75 feet away. Having passed through three solid walls, and the body of the lieutenant governor, the waves arrived at the receiver he'd set up. When intercepted there, a bell rang, a pistol discharged, and a miniature mine exploded (no one was hurt). Bose's transmitter and receiver were both attached to circular metal plates at the top of 20-foot poles, which functioned as antennae.

In his biography of the physicist, Patrick Geddes writes that, inspired by the white flowers offered in Hindu worship, Bose was determined that whatever offerings his life could make should be untainted by any considerations of personal advantage. He invented a device called a "coherer," which did the tricky job of receiving and interpreting electromagnetic signals. He never made a secret of its design, nor patented it. Without it, Guglielmo Marconi couldn't have invented the long-distance radio system for which he is so well known, a deserved level of recognition that has eluded Bose, because without his work in using electromagnetic waves to move information, we wouldn't today have our phones that can send and receive data.

Another world-altering communications technology, the world wide web, which we connect into to use the internet, was invented in the offices of CERN, the European Council for Nuclear Research. Scientists from all over the world work for this organization, and Sir Tim Berners-Lee wanted to create a more effective way to share data, so it could be easily and quickly accessed by the full team. I personally have been fascinated by CERN since I was a teenage physics nerd, but for a slightly different reason: the Large Hadron Collider, or LHC.

The LHC is the world's largest particle accelerator. It is an immense tunnel, 27 km in circumference, shaped like a ring, that sits below the ground in France and Switzerland. Scientists are investigating matter and the origins of the universe by studying tiny particles that make up atoms, and creating explosive collisions between beams of these particles.

Since these particles have a charge, they feel forces from magnetic fields. Using 9,593 electromagnets, charged particles are shaped in two beams, traveling in each direction, around the tunnel. By steadily increasing the strength of the magnets, the particles reach speeds that are very close to the speed of light, and then their path is tweaked to create a collision. Researchers hope that these high-speed, high-energy interactions between particles will answer some really fundamental questions about our origins.

Like the path of these particles, we seem to have come full circle in our understanding of magnetism: from discovering it in the earth, to creating our own, to then inventing electromagnets and using them to learn more about our existence. After a slow uptake, magnets now live by their hundreds in our homes: in discs, memory chips, internet ports, washing machines, telephones, radios, clocks, and meters. But we haven't stopped researching and creating new forms of magnet. In 2022, Dr. Masato Sagawa was awarded the Queen Elizabeth Prize for Engineering for the discovery, development, and commercialization of the world's most powerful permanent magnet.

By using a unique manufacturing technique called sintering (which means heating up and compacting a combination of rare-earth materials) to bring together iron and neodymium with boron (Nd-Fe-B), Dr. Sagawa succeeded in almost doubling the performance of the previous best magnet available on the mass market. His work means that permanent magnets have, once again, seen a step change in size and power, and his innovation is already being used in robotics, domestic appliances, mobile phone speakers, electric motors (including in electric vehicles), and generators for wind turbines. The hope is that the magic of magnets will now navigate us toward a greener future.

Lens

Dear Zarya,

In some ways, I feel lucky. I got to see you before you went into
my body. You were a small blob of different shades of gray, made
up of only 150 cells, which I could see thanks to a highly magnified
pixelated image a doctor printed out for me. Truly, I'm lucky that
you exist at all. Blockages in my body meant that we couldn't
have made you without the marvel that is modern medicine. But
I also feel unlucky, deeply unlucky—why did I have to have those
blockages in the first place? Why was it so, so difficult to make you?

After some months of unsuccessful trying, we went to the doctor
to investigate if there was a problem. The tests started off simply
enough: a blood test, a urine test, then a painless scan. Everything
looked great. More time, more nothingness. Eventually, I was
sent to the hospital for a procedure to look around the organs
in my abdomen, to see if there was tissue from my womb where
it shouldn't be, and to check if the eggs from my ovaries could
complete their intended journey through the fallopian tubes.

While I was unconscious, the surgeon made two small cuts—

one on my left side and one inside my belly button—through which she carefully inserted a thin, black tube: a bundle of special cables, called a fiber-optic cable, that transmits light. It had a tiny camera and light at the end and projected my insides onto a screen. This gave her the ability to peer inside me without having to cut me open extensively in surgery. Then, she passed a plastic tube through my vagina and pressured some liquid through, watching an image of my uterus created by an X-ray machine. The good news was that my organs were nice and clean: no errant tissue, no endometriosis. But the liquid in my uterus had become stuck. My tubes were blocked. As I was trying to digest her words with a brain still foggy from the anesthetic, the surgeon said, in a matter-of-fact way, "I'll refer you for IVF." I couldn't breathe.

Appa and I went to countless more appointments and had countless more scans and tests; then, one day, about a week after my first book was published, I dove into my first cycle of treatment. After a couple of weeks of tablets, we sat in what is now your room, staring at a syringe and a vial full of hormones, and all I could think was, how am I going to knowingly and willingly stab myself with this needle? But I did it. Hundreds of times. To force multiple eggs to mature in my ovaries simultaneously. To regulate my hormone levels to create a nice, thick lining in my womb. To thin my blood, which supposedly increased my chances of pregnancy.

A few weeks later, we went back to the clinic. They used long needles and cameras and screens to empty out the many fluid-filled follicles in my ovaries, each one the size of a grape, each of which hopefully contained a tiny egg. Appa gave his cells, too. Then we went home. All we could do now was wait.

While I recovered (at least physically) at home, scientists worked meticulously to create the possibility of life in the lab. Wearing blue scrubs, head covered, an embryologist peered down a powerful microscope to study our cells and join them together. The little two-celled zygotes were placed in a special jelly that replicated my womb, to nourish them and encourage them to multiply. Each day, someone would check each group of cells to see how well they were growing, if at all. They made ten zygotes that had the potential to become a baby, of which eight survived and thrived— one of which was you, Zarya. But I'd become ill from the treatment. I had to wait a month before they could put an embryo in my body. Back to the clinic, where they selected one of the strongest, and one of your potential siblings was put in me. But it didn't work.

I don't know who that little bundle of cells would have become, and I'll never know. But you stuck. You clung stubbornly onto my uterine lining and grew. You formed your little heart, which I saw pulsing on a black-and-white screen during one of the many ultrasound scans. I watched you squirming around while doctors tried to capture images of you. After a very anxious nine months, you came into the world. Your Appa managed to steal a picture as you emerged from my body, while you were still attached to me, dripping blood, eyes scrunched in agony, screaming at this loud and bright place you'd been pulled into. And now, life is flying by. I try to capture fleeting moments in pictures. My memories feel muddled and unreliable. Time feels warped by your presence.

Zarya, I am grateful that I have you, even though my journey to you was rocky. Even though my journey after you were born has been so, so very tough, not least because of a global pandemic. Even during all these challenging times, I find hope, inspiration, and amazement; in you, of course, but also in the people that made your life possible. I don't just mean Appa and me, or Nani, Nana, Mausi, and Ajji. I don't just mean the specific embryologist who joined you together, or the doctor who placed you back inside me. I don't just mean the many midwives, nurses, and consultants who checked that you were safe and healthy in my belly. I mean the thousands through history, thanks to whom all the science and

engineering involved in your story exist. And my goodness, there
is so much complex science and engineering responsible for your
creation. Zarya, your mama studied physics and is an engineer, so it
is inevitable that I will tell you stories about these things (and I know
you love stories). So, here's an important one. You wouldn't exist
without a seemingly simple little curved piece of glass, called a lens.
This is for you.

    With all my love,
    Mama

Wonder Woman has the Lasso of Truth. Okoye has a Vibranium
Spear. She-Ra has the Sword of Protection. And just as these
special devices give superheroes extra abilities, the lens gives us
humans a superpower. These curved pieces of glass (or other materials that let light through) allow us to manipulate light and see
things beyond the capability of our eyes. Thanks to the lens, the
billions of us that don't have perfect eyesight can see clearly. But
the lens opens much more to us. Because of the lens, we can look
at the insides of minuscule individual cells, at inaccessible geog-
raphies like the deep ocean, and at massive galaxies outside our
own. Because of the lens, we can do incredible things, like study
the origins of the universe or make embryos. But before we get into
the details of lenses, I'll revel in some physics nerdery, because the
story of the lens is inextricably linked to that of light.

    For a long time, light was an enigma. The ancient Egyptians
believed that their sun god, Ra, shone light on them through his
open eye, and, when he closed his eye, it became dark. The ancient
Indians postulated that light was made from infinitesimal particles traveling at an inconceivable speed that radiated outward in
straight lines. The ancient Greeks came up with all sorts of theories about how we could see: because the air between the object

and our eyes is somehow stamped with the object's image; because particles of material flowed from the object to our eyes (with other particles then filling the space, ensuring the object didn't shrink), or because our eyes emitted rays that touched objects. The last theory, which is of course wrong, nonetheless stuck around for a millennium.

Despite this incomprehension, our ancestors realized some of the potential of a curved piece of glass. In an Assyrian palace, now in modern-day Iraq, a Victorian archaeologist discovered a small, circular piece of crystal that had been carefully ground and polished so that one face was flat, while the other was slightly curved. Known as the Nimrud lens, it has been dated to the seventh century BCE, and is thought by some to be an early magnifying lens. The ancient Greek play *The Clouds*, written by Aristophanes in 424 BCE, mentions a "burning-glass" used to concentrate the light of the Sun to create a hot spot. A few hundred years later, Pliny the Elder from the Roman Empire also mentions this device.

The Greeks laid down some basic rules of how light reflects off mirrors and even bends through lenses. But without a full scientific understanding of what light is and how our eyes work, we certainly couldn't even dream of looking at the surface of the Moon as if close enough to touch, or injecting cells into each other with the hope of creating a baby. For the next thousand years or so, erratic and erroneous theories in the field of optics (the study of the behavior of light) continued to hold sway. But in the eleventh century CE, our eyes were opened by the Arab polymath Ibn al-Haytham (known as Alhazen in the West), now regarded as the father of optics, and one of the greatest minds in the world.

Ibn al-Haytham was born in Basra, southern Iraq, in 965 CE. He received a good education and showed great talent in mathematics and science, gaining a reputation for his genius. He created a design for a large dam to tame the River Nile in Egypt, a region plagued by devastating cycles of flooding and drought. News of this idea came to the attention of the caliph Al-Hākim bi'amr Illāh (985–1021 CE), who invited Ibn al-Haytham to go to Aswan and build the dam. It didn't take long for Ibn al-Haytham to become overwhelmed by this huge undertaking, and, apparently in fear for his life, he feigned madness. As a result, he was placed in a mental asylum where he remained for a decade, until Al-Hākim disappeared and was presumed dead.

Such a long isolation had, however, given Ibn al-Haytham time to think and write, and, once released, he published a lot of work quickly. Arguing against the ancient Greek theories, he finally explained correctly how sight works, using simple logic. If the eye emitted light rays that hit an object, in order to see, that object must send some rays back to the eye. But in that case, why are rays needed from the eye? Instead, he proved that it was light from light sources that illuminated objects and then traveled to the eye. He did this by setting up an experiment in a dark room. The light of two lanterns outside the room was projected through a hole, which created two spots of light inside the room. When he covered one light, the corresponding spot of light disappeared. With this experiment, he demonstrated how the camera obscura works, and also did the math, explaining why its images appeared inverted (which I'll touch on later).

Ibn al-Haytham's work related to optics was groundbreaking for many reasons. For the first time, someone suggested, correctly, that light exists independently of vision. This may seem obvious to us now, but at the time, people believed that sight happened instantaneously only because of our eyes; it was alien to consider that light existed separately in nature. Ibn al-Haytham said that this light had a finite speed, and that this speed varied in different materials. We know now that this is why light bends through a lens (although he got the reason why this happens wrong). He also said that light travels in rays along straight lines, and these rays are not modified by other rays that cross their path. For the first time, he conducted a scientific study of images formed by lenses, which often suffered from blurring, which we'll see is something that is bad for microscopes. In short, because Ibn al-Haytham separated light rays from the physiological act of seeing, he succeeded in creating mathematical and scientific models to explain how light and lenses work. He laid the foundation for scientists after him—including Newton, who published his works 700 years later—not only to study and explain light even further but also to engineer spectacles, microscopes, telescopes, cameras, and more. (In another interesting link, physicist Jim Al-Khalili writes that Ibn al-Haytham's discussions on perspective—which were translated into Italian in the fourteenth century—enabled Renaissance artists to create the illusion of three-dimensional depth in their work.)

Much of Ibn al-Haytham's optical theory is contained in his seven-volume Kitāb al-Manāthir (Book of Optics), which was completed in 1017 CE. It's a pivotal work, and not just for its optical insights. In it, he laid the foundation for what we now describe as scientific method: observing natural phenomena, coming up with theories, testing them experimentally, critically reviewing past work, and analyzing the results. His religion, Islam, which encouraged the search for knowledge, inspired him to deepen his understanding of the world, while using the deeply meaningful terms ikhtibar (experimentation) and mukhtabir (experimenter) as his basis. He questioned every theory he came across from the ancients and tested them with experiments, leading to his seminal work. He wrote: "The duty of man who investigates the writing of scientists, if learning the truth is his goal, is to make himself an enemy of all that he reads."

In the subsequent centuries, scientists built on Ibn al-Haytham's foundations. Today, we have complex theories of what light is: that light is formed from waves; that it's formed from tiny particles called photons; that it is a bit of both. I'm going to use the wave theory of light for ease, which says that light is a wave that carries energy and travels through space because of the vibrations or oscillations of electric fields and magnetic fields. (And, as we saw in the previous chapter, these two fields are interlinked.)

When light is traveling in vacuum, far away from other things, there are no (significant) external electric or magnetic fields affecting it. But when it travels through matter, the electric field of light is affected by the electric field inside the material, which effectively applies a brake to the light's field. The light slows down and changes path, and this bending of light is called refraction.

If you shine a narrow beam of light on a rectangular glass block, the light slows down and bends as it enters the glass. Then, as it reemerges, it speeds up and bends back to its original angle. Lenses, with their curved surfaces, force rays of light to do something more interesting. Across the width of the beam, each ray hits a different part of the curve, which means its angle of entry and therefore how it refracts is different from its neighbor. Rather than ending up with a parallel beam of light emerging from the other side as

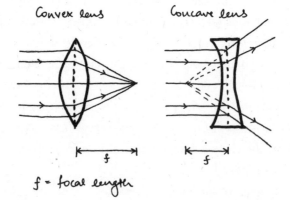

Light rays traveling through convex and concave lenses

with the rectangular block, this changing curvature changes the shape of the beam. Convex lenses, where both faces bulge outward, focus the beam to a point, while concave lenses, whose faces curve inward, spread it out. So, the faces of a lens are designed to manipulate light in a way that allows us to see what we want to see.

The reason lenses change what we see is because of how our eyes (and brains) interpret images. If you look at an ant through a magnifying glass, the light coming from the ground, and the ant, is entering the microscope as parallel rays. The rays refract through the lens and converge, and this is what our eyes pick up. But then, the eyes and brain form an optical illusion. They project the converging beam, as if it had never bent, to create a virtual image that is farther back than the ant, and larger, so we can see it. The ant appears blurry when the rays don't quite converge to a point at the back of our eyes, so we move the magnifying glass back and forth until the ant is in focus.

So, to create a lens for a particular purpose, we need to know how a light ray will bend through it. When you go to your optician and they test different lenses while asking you to read out letters of diminishing size, what they're doing is trying to find the lens that, based on the unique shape of your eye, will bend the light so it focuses on the back of your eye, the retina. The behavior of light in a lens depends on the angle at which the light enters the material, and the characteristics of the material itself. One of Ibn al-Haytham's predecessors in the Islamic Golden Age of science, Abu Sa'd al-Alaa Ibn Sahl, wrote a book around 984 CE called *On the Burning Instruments*, about how to use mirrors and lenses to focus the Sun's rays and create heat. Although the phenomenon had been observed by Aristophanes and Pliny the Elder, this was the first serious mathematical study of using lenses in this way. As with magnets—and, indeed, many discoveries—knowledge and use of lenses far preceded the understanding of them. More than

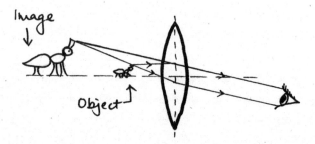

How a magnifying glass creates an image for our eyes

half a millennium later, in 1621, the Dutch astronomer Willibrord
Snellius would figure out a formula for refraction that tells us how
much light will bend between materials (now known as Snell's
Law). But Ibn Sahl had already conceived a geometric formula to
correctly predict how light would refract through a given material
for a given angle of entry. Ibn Sahl's work, however, only came to
light in the 1990s, when the Egyptian historian Roshdi Rashed
pieced together his manuscript, part of which was found in Damascus and part of which was found in Tehran.

The science of optics advanced significantly in the Islamic
empires, but the practical applications of lenses remained largely
limited to burning glasses and simple magnification. Centuries
later, when the Islamic Golden Age of science began to dim in the
Middle East, and as light began to break through the Dark Ages in
the West, Europe's Renaissance thinkers built on the work of their
medieval counterparts to truly harness the superpower of the lens.

The mention of microscopy in that era conjures up images of
wigged men peering into gilded tubes mounted on adjustable
arms, or shiny brass instruments with narrow cylinders, intricate
screws, and thin plates affixed to solid bases. But it was actually a
much cruder-looking device that revealed a new world.

Robert Hooke, the polymath whom we met in the chapter on
springs, is famous for his book, *Micrographia: or Some Physiological*

Descriptions of Minute Bodies Made by Magnifying Glasses, With Observations and Inquiries Thereupon, which he published in 1665. Hooke busied himself with making familiar objects larger, and meticulously drawing them, such as his famous, beautifully shaded, black-and-white sketch of a flea. In the preface of this work is a description of one of these simpler forms of a microscope: small, hand-held, with a single, hand-ground lens. No doubt inspired by Hooke's work, a Dutch shopkeeper with little formal education decided to look closer, leading him to see many things that humans had never seen before.

Antony Leeuwenhoek set up a business as a draper in 1654, aged twenty-two. A loner and obsessive enthusiast, he was fascinated with the natural world, but remained in Delft, trading material and holding minor government jobs. He wasn't well-educated, at least in comparison to the other scientists or philosophers of his time; he didn't speak Latin. Leeuwenhoek created simple magnifying glasses in order to verify the quality of his cloths by checking the thread counts. He didn't do much with his lenses until he was forty, when he began making his own microscopes to study nature. Over the next fifty years, he went on to construct nearly 500 single-lens microscopes, but he was highly secretive about how he formed his lenses.

I saw one of his constructions preserved at the Science Museum in London. It looks nothing like the elaborately constructed and curlicued microscopes of the Renaissance; indeed, at first glance, with its flat, rectangular riveted brass plates surmounted by a long screw, it looks more like an obscure lock mechanism than a microscope. But sandwiched between the plates is a single hole, and within that hole sits a lens. Leeuwenhoek's microscopes were small: the lenses, typically polished from glass spheres, were just 1 mm wide, and the metallic plates about 4–5 cm. At the bottom of the vertical plates is an L-shaped bracket, through the horizontal leg of which extends the long, threaded screw, restrained at the top by a metal block.

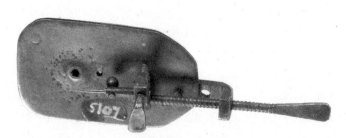

Leeuwenhoek's microscope

Leeuwenhoek placed his samples in a small glass vial, which he attached to the top of the large screw with glue or wax, with a couple more screws allowing him to tweak its alignment. Holding the microscope up to his eye, he could peer through his handmade lenses, some of which could magnify objects by an astonishing 266 times. To put this in perspective, the microscopes with two lenses invented in the late sixteenth century by the Dutch father and son team, Hans and Zacharias Janssen, could only magnify up to a maximum of ten times, because of the limited quality of the lenses and the blurring caused by the effects first studied by Ibn al-Haytham.

Leeuwenhoek's microscopes led him to make some of the most startling discoveries of the seventeenth century. In 1674, he wrote in awe about observing beautiful, spirally wound green streaks, small green globules and little "animalcules" with diverse colors and glittering little scales—microorganisms that today we call protozoa, ciliates, and algae. He pricked his finger and was the first to see red blood cells. Then he discovered bacteria. An illness the previous winter had temporarily caused the loss of his sense of taste, and on inspection, Leeuwenhoek found a fuzzy layer on his tongue. This inspired him to look at an ox tongue through his microscope, where he saw what we now know are taste buds. He then wondered how these little projections interacted with potent tastes, like pepper and ginger, so he examined infusions of spices. He must have been shocked to see thousands of tiny eel-like organisms, which were in fact bacteria. This was, for humans, a life-changing discovery. Even though it took about 200 years before scientists figured out that these bacteria caused a whole host of illnesses, without his work we wouldn't be able to treat multiple diseases with antibiotics that save countless lives.

The discovery he made that relates to Zarya's inception took place when he studied one of his bodily secretions. In contrast to the dozens of other usually colorful and descriptive letters Leeuwenhoek wrote to the Royal Society in London, his 1677 account of his findings is uncharacteristically sheepish, presumably due to concern that it might cause disgust or discredit among learned men. So, he provided assurances that the sample was obtained from what nature leaves behind after conjugal coitus, without any sinful defilement on his part. In these remains, and in the "seed" of a multitude of living animals, he saw tiny animals—so tiny that he believed even a million of them wouldn't be as big as a grain of sand. They had round bodies with a front part that was blunt and a back part that was pointed, with a thin tail that helped them move like eels. He had seen sperm.

Combined with the theory that all female animals have eggs, which also made its appearance in the mid-1670s, it's surprising that another 200 years went by before humans finally understood how babies were made. It boggles my mind that we had made such progress in all sorts of science and engineering by the nineteenth century, but still believed that it was only the ovum (if you were an ovist) or the sperm (if you were a spermist) that created new beings (even if we accepted that the sperm and egg needed to interact in some way). Finally, in 1875, after spending hours studying things through his microscope, the German biologist Oskar Hertwig saw a single spermatozoon entering an ovum of a sea urchin, leading to cell division. Fertilization. And that's when we fully understood—and proved—how babies are made.

Still, in the early 1900s, doctors didn't really know the details of the cycle of ovulation. Without modern ultrasound scanning machines that produce those familiar black-and-white images of fetuses, they couldn't predict when an egg would be released by an ovary. Little was understood about where it was fertilized by the sperm, and how long it took for cells to multiply to create an embryo, and where the embryo implanted itself. Clinician John Rock and Harvard pathologist Arthur Hertig worked together in the United States for nearly two decades from the 1930s, producing groundbreaking work in tracking the early development stages of human embryos. At the same time, Rock hired the scientist Miriam Menkin to work on creating human embryos outside the body, or in vitro. Their work built on years of research, which included Samuel Leopold Schenk's report in 1878 of the in vitro fertilization of the eggs of rabbits and guinea pigs, and Walter Heape's success in transferring embryos from one breed of rabbit to another in 1890.

Menkin had hoped to get a PhD in biology, but ended up having to financially support her husband through medical school and having to care for their two children. Instead of pursuing her own research, she assisted others, and with experience with IVF in rabbits, she brought her knowledge and meticulous scientific skills to Rock's laboratory. She called herself Rock's "egg chaser," because every week, when one of the study volunteers was having surgery, she would stand outside the operating theater in the basement of her lab, waiting to see if any ovarian tissue was removed, and then run up three flights of stairs to look for eggs. It was tedious work: nearly a thousand women agreed to be a part of the study, and Menkin found eggs in the samples of only forty-seven of them. And finding the egg was only the first step; then, she had to try and fertilize them.

After six years of this energy-sapping weekly routine, on February 6, 1944, while dozing at her microscope, Menkin woke up to see two cells. She had been awake the night before with her teething child, and because of the exhaustion, hadn't washed the sperm as many times as usual. She'd also used a more concentrated sperm mix, and left the egg and sperm in the dish for longer than usual after she lost track of time. In the excitement of relaying her work to her colleagues—and a heated discussion on how best to preserve the zygote—she forgot to photograph the cells. But she succeeded in fertilizing three more eggs that year (which were carefully photographed).

The microscope Menkin used wasn't particularly powerful: it magnified the smallest human cell, the sperm cell, by thirty-five times, which was enough to see it and watch it interact with the largest human cell, the egg. This magnification was also sufficient to watch the cells divide to make sure the embryo was growing well. Her microscope was a conventional one; it had a surface on which the dish with the cells was placed, with an eyepiece and light source overlooking it. Light reflected off the cells into the magnifying lens, then moved through the eyepiece into her eyes. Her proof that human embryos could be created in a lab was instrumental in furthering the treatment of infertility—but to actually treat infertility, the embryo needed to be nurtured so it could grow outside the body for a few days, then be put back inside the womb, successfully implant and not miscarry. Working through all this complex science meant it took another thirty-four years before Louise Brown, the first IVF baby, was born.

The microscope used to create embryos now is rather more powerful and complex than Menkin's. To give us the best chance of success, the embryologists suggested that they use an intricate procedure called ICSI (intracytoplasmic sperm injection). Rather than mixing up lots of sperm cells with an egg in a dish, in ICSI, a single sperm cell is injected into the egg. A micromanipulation system—a system of small, movable arms with tiny needles—is needed to enable human hands to control the tube that holds the egg on one side, and the needle that injects the sperm with the other. Our fingers simply couldn't hold the tube and needle by themselves; after all, they are made from glass, and the injecting needle is just 0.005 mm thick (the tube that holds the egg is only slightly thicker). It wasn't surprising to hear from Christiana Antoniadou Stylianou, the embryologist I spoke to, that she held her breath every time she did an ICSI procedure for the first five years of her career. Clearly, some amazing skill is needed, but it couldn't be done at all if Christiana couldn't see what she is doing. The reason this procedure is so modern is because you need to be able to observe the egg in three dimensions and roll it around in order to identify a tiny structure within it that is crucial in correctly distributing chromosomes in the dividing cells. If this structure gets pierced by the injecting needle, the egg dies—no embryo. To see this structure, the eggs need to be magnified around 400 times, ten times more than Menkin's microscope could have done. The microscopes used by embryologists today are called inverted microscopes. In a conventional microscope, reflected light is collected by the lens. This means it loses some intensity and doesn't allow for distinguishing between different layers of the medium that holds the cells. Inverted microscopes are arranged so the lens collects light that passes through the medium in the dish, giving the embryologist a clearer view and the ability to change the focus of the lens so they can see the layers of the medium and also see the egg as a sphere. The technology needed to make ICSI possible—the microscope, the manipulation system, the needles, and the research to ensure ICSI babies will remain healthy—only came together recently, with the first baby born in 1992.

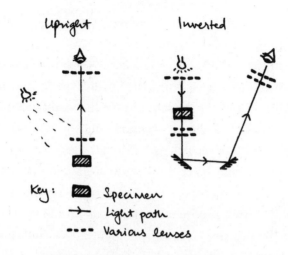

Simplified sketch of how light travels
through upright and inverted microscopes

The story of the science and engineering that created my daughter feels like it was over a thousand years long, but it also feels scarily short. Leaps in treatments and techniques have been made within my own lifetime. The oldest IVF humans are only five years older than me, so we don't even know what will happen to them in middle age. Maybe they'll suddenly mutate and acquire some exciting superhero powers. Maybe not. But one thing is for sure—without this convoluted journey through the science of light, lenses, and semen samples, Zarya wouldn't be here.

Lenses revealed the world of the microscopic by letting us see things that were too small for our eyes, giving people like me the chance to become pregnant. That's just one example, of course; the microscope meant we could study bacteria and viruses, and thereby save lives. At the other end of the scale, lenses in telescopes expanded our view by bringing into focus things that were too far away for our eyes. From believing that the Earth is flat to the realization that it is but an insignificant dot, one of trillions, spread out over inconceivable distances, humans have come a long way in understanding our place in the universe.

Leeuwenhoek left an enduring legacy of looking at little living organisms through a lens. When the Covid-19 pandemic hit, scientists swiftly mapped out the structure of the virus and then raced to create vaccines, so eagerly awaited, and these simply couldn't have been created without lenses. To battle cancer, another devastating disease that touches so many of us, requires looking as closely as we can at these cells that can carry on replicating forever. But not only are we looking at cancer cells in a highly magnified static view, we are also tracking—in real time—what's happening inside them. We can do this by shining a special type of light on the cells: the laser.

A laser (the name is an acronym for Light Amplification by Stimulated Emission of Radiation) is an artificially created beam of very pure light. It is extremely powerful—it can cut through metal and pierce diamonds. Laser beams are quite different from the light that comes out of a lightbulb or a flashlight. These more familiar light sources give off white light—a mixture of light waves of all different colors or wavelengths—but lasers are made up of a single wavelength or a very narrow range of wavelengths. The light from a flashlight spreads out into a cone and dissipates, but lasers have a tight and narrow beam that's near parallel, so it travels consistently over much longer distances. The waves of light coming from the flashlight are mixed up (like people walking around randomly on a street), but the laser waves are synchronized or coherent (like soldiers marching in step). Another special thing lasers can do is form extremely short pulses in contrast to the continuous waves of light emanating from a flashlight.

The lens isn't necessarily needed to generate a laser emission. But it—along with its triangular cousin, the prism, and the mirror—is essential for transporting and tweaking the emissions, making sure that the beam coming out of the machine has the right width, intensity, and level of synchronization for your purpose, turning laser emissions into practical tools.

Some of the most advanced and powerful lasers in the world even rely on something that's a cross between a lens and a mirror to generate the emissions in the first place. The titanium:sapphire laser is one that produces some of the shortest and most powerful laser pulses. To generate these pulses, a green laser is shone onto a titanium:sapphire crystal to excite its atoms. These excited atoms emit pulses of light that are of a longer wavelength (red). A pair of these mirror/lens devices are arranged in such a way that the green light from the laser passes through them and is focused down to create a high intensity point at the crystal, behaving like a lens. The red light pulses that are emitted from the crystal, however, gets reflected back and forth between the mirrors until they form a strong beam, and is then allowed to pass through into the main instrument.

Professor Stanley Botchway uses one of these lasers, which can produce pulses as short as six femtoseconds (or six quadrillionths of a second), as a form of microscope for his research. Stan is investigating how the DNA inside our cells gets damaged when exposed to different types of radiation. At the moment, broadly speaking, radiotherapy to treat cancer fires high-energy X-rays at tumors to kill the cancerous cells—but the problem is that healthy cells around the tumor are also damaged, causing difficult side effects.

One of his projects is testing how to use lasers alongside medication, a process called photodynamic therapy. The medicinal molecules are administered to the patient in an unactivated form, and they seek out cancer cells because they have been designed to prefer environments with less oxygen (tumors grow so fast that the blood vessels don't have time to catch up, so there's less oxygen). The drug targets the skeleton of the cancer cells that holds the cells together. Then, the titanium:sapphire laser is carefully pointed at the tumor. When hit by the femtosecond pulses, the drug becomes activated and kills the cancerous cells.

Since the surrounding healthy cells will not have attracted much of the drug, and also because the laser can be so carefully aimed at the tumor, the healthy cells are minimally affected. The red light from these lasers can penetrate deeper into tissue than X-rays, so it is far more effective in treating hard-to-reach tumors while reducing the side effects.

It's worth noting that the cells used in this form of research into cancer, as well as countless other diseases, are called HeLa cells. These are cancer cells that were taken for research without consent from a Black American woman, Henrietta Lacks, who died at the age of thirty-one in 1951. The cells taken from her have been alive and replicated ever since, and are one of the most important cell lines in medical research. Johns Hopkins issued an apology over fifty years later, but her family have received minimal compensation.

Scientists are also using these lasers with incredibly short pulses to watch biological processes that happen inside cells. There are interactions or changes that occur in our cells because of disease—and also in healthy cells—so quickly that it's impossible to see with normal microscopes. By sending these pulses that are a tiny fraction of a second to the cell and taking hundreds of thousands of images every second, we get to watch the building blocks of life in action.

There is also a different way in which lenses give us fresh insight: through the medium of photography, which captures images that our eyes, in theory, are capable of seeing, but in practice, we cannot access. At the center of the camera is the lens. Of course, technically, a camera doesn't need one: images can be made through pinhole cameras without lenses. But I've selected the camera because adding the lens changed photography—and, subsequently, society. The quest for the perfectly focused, clear photograph in a range of different settings drove huge innovation in the design of lenses. Without them, we wouldn't be able to take sharp images of small things and large things, of still things and fast-moving things, of close-by and faraway things. One of the greatest portrait photographers of the twentieth century, Armenian-Canadian Yousaf Karsh, said, "Look and think before opening the shutter. The heart and mind are the true lens of the camera." I find this a striking comparison; by likening a photographer's heart and mind to the lens of the camera, he identified the lens as the camera's soul.

Thanks to cameras, fleeting moments from history have been preserved for us to see today. They've given us defining images of inaccessible places and people on our own planet. Many of these have sparked or spread social change. I remember when the photograph taken by Justin Hofman of a tiny seahorse with its tail wrapped around a waterlogged plastic swab went viral, bringing to light the devastating impact our waste can have on nature. Photojournalism in 1960s Vietnam—most famously Nick Ut's photo of Phan Thi Kim Phuc (often simply called the "Napalm Girl")—shattered the false reality that ordinary Americans were living. It showed them the real face of war, and played a role in the protests that followed. India's first female photojournalist, Homai Vyarawalla, captured inspiring moments between ordinary people and political leaders, and shared them widely during the Independence movement in the 1930s and 1940s; her work has now become an important record of that era. (At first, because she was a woman and unknown, she had to publish her photos under her husband's name, which paints a disappointing picture, but Vyarawalla turned this to her advantage: the fact that people didn't take her seriously as a journalist allowed her to "go to sensitive areas" and take pictures that others couldn't have captured.) Moreover, it's not just those behind the lens who have enacted social change. Frederick Douglass, a formerly enslaved man turned abolitionist who was known for his rousing speeches, consciously used the "democratic art" of photography to counteract the derisive and vile caricatures of Black people that were common in his day, and instead impart the true humanity and diversity of his race. Douglass, not Abraham Lincoln, was the most-photographed American of the nineteenth century.

While it might be a democratic art with the power to educate and change minds, the history of photography is nevertheless filled with ambiguities. At the beginning, not all people were equal before the lens. To create true portraits of Douglass required change. The earliest cameras were designed by and for people with light skin, leaving images of Black people flat and lacking in detail. To do justice to all humans with our wide variety of skin tones, more light needed to travel through the lens and be captured on the film. Cameras were also used in many countries by colonizers to exert power over the colonized and share with those back home images of people considered to be inferior or less evolved. These images make for deeply uncomfortable viewing: one shows people from a remote tribe of the Andaman Islands, naked, being choreographed into poses with a white man standing over them, referred to as their "keeper." Women and men who covered their faces for religious or spiritual reasons were forced to expose them so that records could be made against their will. The lens has given us the power to see more than we could ever have imagined, but as the superhero Spider-Man was warned: with great power comes great responsibility.

Camera technology and the ways in which we photograph have developed rapidly over the last century, and even during my own lifetime. The thousands of photos that I've taken of Zarya are saved in some nebulous cloud as zeros and ones; there are already more of them than there are photos of me from my entire childhood. The pictures of me were created on photographic paper from negatives on film. Now these photos have a retro look about them, with slightly worn edges, slightly bleached colors. I like leafing through them when I visit my childhood home, slightly incredulous at the clothes I wore back then: matching blue-and-red checkered pants and a sweater, paired with red suspenders. Even most of my memories from my university are preserved on printed sheets. In my early twenties, I finally bought a digital camera. Now I carry a smartphone that has a far more powerful camera than my digital one, and I take multiple photos a day. This year, globally, we will take over 1.4 trillion photographs.

The predecessor to the camera, the camera obscura, didn't have a lens until the sixteenth century. Camera obscura is Latin for "dark room," and camera obscuras were just that: dark rooms that you went inside to look at an image of the outside. The image, which was upside down and reversed, was formed thanks to a small hole in one wall that let in a small beam of light (this was how Ibn al-Haytham set up his two-lamp experiment). Cross your arms in front of you at your wrists. Your forearms and hands are beams of light, and your wrists represent the pinhole. You'll see that because your arms are crossed, your right hand ends up on the left, and your left hand on the right. If you manage to contort yourself so that your right elbow is directly above the left, again you'll see that the "image" between your hands is inverted, with the right hand below the left.

The issue with the camera obscura was that the images were not very bright or sharp. To make an image brighter, you'd need to expand the hole to let more light in, but then the light rays could crisscross anywhere around the large hole; without a sharp point of entry, the image would be even more blurred. As lens-making progressed in sixteenth-century Europe, a lens was introduced at the aperture of the camera obscura. Now the hole could be made larger to allow more light to enter, because the lens would bend the rays, sending them to defined points to create a sharp image. Still, these devices required you to physically walk inside them to see an image of whatever was outside—not exactly anything new or horizon-expanding—and so they were largely used for entertainment. You couldn't capture the image, except by tracing over it manually. Artists employed the camera obscura to create projections of three-dimensional scenes or objects, which they would then paint on a flat surface. (In his book *Secret Knowledge: Rediscovering the Lost Techniques of the Old Masters*, the artist David Hockney put forward the intensively debated idea that the increased sophistication of Western art from the fifteenth century onward is in part due to the use of the camera obscura and other optical devices—an idea that he has since explored further with the physicist Charles M. Falco, and is now known as the "Hockney-Falco thesis." Falco has also argued that it is Ibn Al-Haytham's Book of Optics that provided the original inspiration for this artistic development.)

While camera technology remained in its infancy, lenses of increasingly better quality were being made for microscopes and telescopes. Scientists were struggling with the images being created by lenses in the sixteenth and seventeenth centuries because of issues called aberrations. Rainbows show us that light is made up of a mixture of different colors. Each color has a different wavelength, which is the distance from the peak of its wave to the next trough. One problem—called chromatic aberration—arose because the amount light bends through a lens depends on its wavelength, so red light bends slightly differently to blue light. The other problem was that, although in theory light rays passing through a convex lens should focus down to a single point, in reality there are tiny differences in how the light bends when it's close to the center of the lens compared to when it's farther out—a phenomenon called spherical aberration (which is something that Ibn al-Haytham studied in detail). Added to these aberrations was the fact that lenses were being made by manually grinding and polishing blobs of glass. While instrument-makers were highly skilled in their craft, they couldn't create lenses that achieved the perfect theoretical shapes needed. All these factors caused images to lose sharpness, which in turn limited how much objects could be magnified: at some point, the image would become so blurred that you couldn't see anything useful.

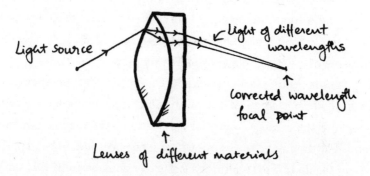

Using lenses made from different materials to reduce aberration

The issue of aberrations was solved in the eighteenth century by telescope designers. They took two different types of glass made into differently shaped lenses and fitted them together to create a so-called achromatic lens or doublet. The aberrations caused in each of these lenses were different because of their material and their shape. But when combined, the aberrations got canceled out, creating a better, sharper picture. This allowed scientists to magnify small or faraway objects to a much greater degree than ever before.

Although the design of lenses marched forward, chemistry hadn't quite caught up: we hadn't really worked out how to imprint light on a material permanently. In the early nineteenth century, the real power of the lens in photography was unleashed with the introduction of a plate covered in chemicals placed behind the lens. Finally, we could walk away from a camera with a captured image. This was the birth of photography as we know it today.

The story of the camera then became about the delicate balancing act between lens technology and film technology, which defined what sort of images we could capture. If you've ever wondered why photographs of people from a couple of hundred years ago look so stiff and unsmiling, it's because they had to sit in a chair— which had a torturous-looking contraption to hold their heads in place—for a long time. One of the first commercially successful processes of creating images was the daguerreotype, named after its inventor, Louis-Jacques-Mandé Daguerre. The photographer put a sheet of copper coated with silver and exposed to iodine vapor into the camera, a largish device that was held steady on a stand. For the image to get imprinted, the sheet had to be exposed to light for up to thirty minutes (later versions brought this down to around a minute), after which it was exposed to mercury vapor to bring out the image. The image was then sprinkled with salt to make it permanent. A number of the chemicals involved were extremely dangerous (mercury poisoning can damage the nervous system, causing delirium, changes in personality, and memory loss—symptoms sometimes referred to as "mad hatter syndrome," as it was prevalent in the hat-making profession because of the widespread use of mercury to treat felt), so photography remained in the hands of professionals until the twentieth century.

The lenses used in portrait photography were designed for a narrow field of vision; after all, the aim was to create a sharp image of one person fairly close to the lens. Lenses with a different shape were needed to capture landscapes, where focus was less important than a broad field of view. There was even a lens made from a hollow sphere filled with water, which gave a wide (if distorted) view, designed to take panoramic images. Portraits and landscapes were pretty much all that could be photographed in those days because of the long exposure times and the relative bulk of the cameras themselves.

In the 1890s, new types of glass arrived on the market. One, called barium glass, refracted light by a relatively large amount because of its molecular makeup, while another was an improved version of so-called flint glass that in contrast to barium glass, had low refraction. At the same time, rather than relying on trial and error, lens-makers developed mathematical equations to predict how lenses would work based on their curvature. The doublet lenses formed from these materials and shaped by science led to high-quality, larger lenses that let lots of light into the camera, helping speed up exposure times.

From the late nineteenth century onward, lenses in cameras rapidly became increasingly complex. Doublet lenses became triplets (with three layers of glass); such lenses were then arranged in arrays of three or four or even more, all designed to carefully cancel out as much aberration as possible. Even though lenses are made from transparent materials, which in theory transmit most light, they do reflect some. In the 1930s, the scientist Katherine Burr Blodgett developed coatings for glass that were only a few molecules thick, which maximized the amount of light traveling through the lens—an early antireflective layer like those many of us have on our glasses today. As lenses improved and let in more light film technology developed simultaneously, leading to exposure times dropping from minutes or seconds to small fractions of a second. Today, typical images are taken with an exposure time of around 1/200th of a second, although it's even possible to go as

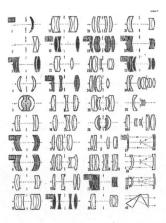

A timeline of lens development. The Focal Encyclopedia
of Photography, *desk edition, London: Focal Press*

short as 1/8,000th of a second. In the early days, photographers
simply uncovered, then re-covered their lenses once the exposure
time was up, but with these new cameras, the manual movement
risked overexposure. That's when mechanical shutters were introduced in front of lenses to reduce the amount of time light was
allowed into the camera. More mechanics were then introduced so
you could move one set of lenses relative to the other, allowing photographers to zoom in and out on their subjects.

The advancement in lens and film design led to cameras becoming smaller, easier to use, cheaper, and less dangerous. In 1888,
there appeared on the market a camera called the Kodak No. 1,
which had one fixed shutter speed, one focal length, and a crude
viewfinder, and was preloaded with film. When the roll of 100
images was full, the entire machine was sent to a factory in Roch-
ester, New York, then reloaded and sent back while the film was
developed. Finally, cameras reached the general public. Family
photo albums, once filled with formal portraits, became records
of spontaneous events. Short exposure times allowed us to capture a moment in time of busy streetscapes and moving people and

objects, bringing life to the images. Cameras were taken overseas, and for the first time, people could experience other countries and cultures from their own homes.

As I've said, one of the biggest changes to photography that happened in my lifetime was the move from taking pictures on film to capturing them digitally. This, initially, didn't have a big impact on the design of lenses, because the electronic sensor that records digital images is designed to be the same size as film. This meant that as we transitioned between the two, the lens attachments for film and digital cameras were interchangeable.

However, our smartphones demand a lot more from lenses. Unlike hand-held cameras, which have space to have a larger lens and greater distance between the lens and the film or sensor, the cameras in our phones have small lenses that are very close to small sensors, so they focus light onto a much tinier area in order to capture an image.

Early phones had one lens with one focal point: that is, the distance at which a photographed object would be very sharp. Then, recently cameras with autofocus were introduced, where tiny little mechanisms physically moved the lens back and forth a bit, to increase the range of focus. Most recently, we are seeing phones that have two or even three lenses. At one extreme, there's a lens that has a very long focal length for photographing distant objects. At the other is a short focal length that can take wide shots, as well as one that does something in the middle. The sensors in our phones that detect the light and turn it into electrical signals, which are then converted into images, are far smaller than those in digital cameras, to keep down the weight and thickness of our phones. A new opportunity to improve these small phone cameras comes with the use of periscope focus, where a prism or angled piece of glass is used to turn the light from the camera lens by ninety degrees, then send the beam along the back of the phone through a number of lenses that can be adjusted.

Over the next few decades, we'll probably see continued development in the materials used to make lenses, making them lighter, clearer, and cheaper. Software will continue to enable us to design really complex arrays of lenses on a computer, rather than trying arrangements out manually like the early lens-makers did. We already have prosthetic lenses that are used to replace our biological ones when they cloud up and create cataracts.

In the last two decades, engineers have developed bionic eyes to help restore the vision of some people affected by age-related macular degeneration, a condition that affects millions worldwide, causing distorted vision and loss of sight. An electronic chip just 2 mm wide is implanted behind the blind eye. The patient wears glasses with a video camera that is connected to a small computer worn on a waistband. The camera transmits data to the computer, where it is processed and sent back to the glasses. The glasses then project a beam of infrared light to the chip through the eye, and the chip sends electrical signals to the brain, which then interprets them and allows the patient to see. Scientists in China are researching materials that respond to light in the same way our pupils do—expanding when dark and shrinking when bright—to replicate a natural reflex in an artificial eye.

Superhero powers to see things we otherwise couldn't see are exciting and educational, and can be beneficial to humanity, but I'm struck by the fact that technology is also being developed for reasons that enhance our individual experiences. Just as cameras have enabled us to look back in time to our past, allowing us to see and experience our precious memories, as our own eye lenses begin to cloud up or fail, future lens technology might just be able to help us see our worlds, in our present, once again.

String

The most rewarding—and disconcerting—day of my career as an engineer was standing for the first time on the solid steel deck of the Northumbria University bridge. Eighteen months earlier, when I started my first full-time job, I was handed designs for this beautiful structure. I marveled at the fact that what was then only a notion would one day become a fully formed, three-dimensional reality. When that day arrived, I traveled to Newcastle (hoping I'd done all the math right) to stand on the steel that I had previously seen only on paper.

Every engineering project has some parameters that define what we can and can't do. Electrical cables must transmit current without shocking us if we touch them. Washing machines must fit through standard door widths, and must not leak. In this case, the engineering team had to maintain a safe height above the motorway for trucks and buses to pass, while also tying into the existing pathways on either side (and also not wobbling or collapsing). This left a limited amount of space in which to include structure. The simplest form of bridge we could have designed was horizontal steel beams supported only at the two ends. At 40 meters long, this bridge was on the small side, relatively speaking. Even so, ensuring it was strong enough and didn't sag too much when walked over would have required extremely deep beams that would have encroached into the clear height above the traffic. One possible solution to this was the addition of a column to support the bridge from underneath, but this would have had to be positioned on the motorway and, even if it was housed in the central reservation, there was always a risk that a collision with the column would render the bridge unsafe. We needed to find a safer, more elegant alternative.

Our solution was to suspend the bridge from above. Walking across the finished bridge, I couldn't resist gazing upward at the six pairs of cables that soared over me. The immense weight of the bridge itself, and of all the people who walked over it, was channeled through those cables, making it stable and robust.

The strong steel cables that suspend our longest bridges—and my short one—are an evolution of a seemingly simple piece of technology: string. String enabled our ancestors to spin out a series of innovations that have defined the lives we live today. We used cord to stitch together animal skins, and later created a variety of different cloths. We made this into clothes to protect us from the elements, allowing us to occupy colder and hotter climates than our skin alone could handle. We have passed stories down the generations through music played on string instruments. We explored the world and colonized countries using ships that needed woven material for their sails and rope from which to swing them. We created canvas from fibers, onto which we painted and recorded our experiences. We continue to stitch together wounds, bridge valleys, and protect our bodies, all thanks to the quiet strength of string. And this is the defining feature of string—you can make it as strong as it needs to be for its task, yet it's flexible.

The natural world offers a number of animals that can create a strong, threadlike substance, such as spiders and silkworms. People possibly drew inspiration from such sources to create their own version. However, it was not humans who first invented string, but the Neanderthals. In the southeast of France, in a valley near the Ardèche River, is a set of caves called the Abri du Maras, where the Neanderthals are known to have lived for long periods during the Middle Paleolithic era (c. 300,000–30,000 years ago). In 2020, it was reported that archaeologists exploring 3 meters below today's ground level, in a layer that is between 41,000 and 52,000 years old, had found a stone tool, to which was stuck a tiny piece of string, just 6.2 mm long and 0.5 mm thick.

This is an exciting discovery, partly because it is far older than the previous oldest example of string (which dated back a mere 19,000 years), and partly because it shatters stereotypes of Neanderthals being our not-so-smart cousins. Being able to make this product means Neanderthals could have created bags, mats, baskets, and even fabric. String may have played an important role in how they led their daily lives, and how they spent their time, since making it is a laborious and time-consuming task.

It also tells us something about their cognitive abilities, because the way they constructed their string required some mechanical understanding. String is made from fibers, but a single strand on its own is fragile and not particularly useful. To render it more serviceable, several strands need to be combined in a way that causes them to rub up against each other; the frictional forces between them are what give string its strength. In the sample found at Abri du Maras, the Neanderthals had twisted fibers extracted from bark to create a yarn. The direction of the twist they used—which is still commonly used today—is known as an "S-twist," because, like the middle section of the letter S, you can see the fibers wrapping from the top left to the bottom right along the length of the yarn. Then, three separate pieces of yarn were twisted to form cord in the opposite direction—known as a Z-twist, with the yarns wrapping from top right to bottom left.

This interweaving of layers in twists that oppose is central to the wonder of string. If you twisted your fibers to create yarn in one direction, and then twisted multiple pieces of yarn in the same direction to strengthen it, you would fail, because a small tug would cause the twists to unwind and the string to stretch out and unravel. Having the layers twisted in opposite directions means that, in order to unravel, they need to untwist in opposite directions, and the friction between these layers of yarn prevents this from happening.

Although they didn't know it at the time, the Neanderthals were replicating biology. Our most important natural fiber today, wool, is built up from multiple complex layers of keratin (a protein that also makes up our nails and hair), and the innermost layers have this same opposing, twisting structure.

Since no stone is left unturned for my readers, while research-ing this chapter I even bought some yarn and had a go at knitting. I started with some chunky yarn that was simply one thick strand with a gentle S-twist. Sure enough, this was easy to untwist on its own, but once knitted into some slippers, the intentional entangle-ment and looping of the yarn created complex frictional forces, and it held strong. Then I tried using what's called worsted or Aran wool, which essentially has the same structure as the Neander-thal string: three strands that have been S-twisted together, then assembled in a Z-twist, making it trickier to unravel. (By this point, I'd graduated to the far more ambitious project of a sweater, and it's fair to say that knitting became a welcome and—I con-vinced myself—justifiable distraction from writing.)

The ancient urge to create a stringlike material perhaps shows how essential and fundamental string in its many forms is in so many aspects of our lives: from holding up bridges to prevent us from plummeting, to protecting and even reshaping our bodies, to producing melodic sounds. I don't often see string featured in lists of "top inventions," but I think it should be. In arguing the case for its inclusion, I'd call up no less a person than the Roman engineer and architect Vitruvius. In his *De Architectura* (which had a huge influence on Renaissance architecture), he identifies three principles of good design: firmitas, utilitas, venustas—strength, utility, and beauty. And string persuasively brings these three strands together.

Firmitas

One of the constant challenges for engineers is designing a structure that is practical, elegant, and unobtrusive. Traditionally, building generally used whatever materials were most readily available, such as stone or bricks or—once we'd figured out how to create it—concrete. Strongly resistant to compression or squashing forces, these materials were serviceable and did the job, but in the past, they led to bridges that were chunky and heavy, requiring many columns to support their weight. Crossing long stretches of water or deep valleys was at best expensive and time-consuming, and at worst, impossible.

Then, along came a different type of bridge. Long pieces of rope were flung across challenging terrain and used as a starting point to create crossings. Twisted rope makes a great structural material when pulled or put in tension. Since rope is made up from many fibers, even if some of them stretch and snap, the rope itself won't fail dramatically without warning. We would see the fray, signs of decay, and know that it needed repair or replacement. These rope structures effectively channeled the pulling forces to their ends, which were securely attached to the rock faces or foundations (a fact I have to remind myself of, as I walk across such a bridge, because even though I'm an engineer, I find the experience terrifying). And in certain situations, it was much easier to make rope and put it in place than to carry out the laborious process of fixing stone or holding concrete while it hardened.

An extraordinary example of this that still exists to this day is the Q'eswachaka Bridge in Peru, whose walkway, handrails, and supports are all woven entirely from natural fibers, giving it the appearance of a supremely ambitious macrame project. Originally built during the Incan Empire more than 500 years ago, the bridge formed part of the network known as the Great Inca Road that connected the empire. For a long time, the 30-meter-long structure was the only connection between two villages on either side of the Apurimac River. Every spring, communities on either side of the valley come together for a ceremony of renewal. Quechua women sit at the top of a canyon and twist ichu grass into long cords. Then, the men braid the cords together to form six immense ropes that are as thick as their thighs: these form the main structure of the bridge. Four are laid side by side across the valley to form the walking surface, and the last two are installed at shoulder height on either side to create the handrails. To complete it, men straddle the four main ropes. Two or three more men stand behind them, bringing over thinner strands of rope that are woven carefully across the six main ropes to stabilize them and create a pathway to keep pedestrians safe from the ravine below. Finally, the old structure is cut loose, and left to tumble into the gorge beneath.

Originally, the Quechua people, like other engineers in the distant past, were making use of the strongest and most practical materials to hand. But in time, other materials became available, and with them new engineering solutions to spanning an obstacle were possible. As we developed our skills in mining and working metals, we created bridges that were hung from chains, usually

made from wrought iron. Some of these were chains in the way
we typically envisage them: links that are connected together in
a long sequence. More often, the chains were made from flat bars
of metal that were joined together using round metal stubs called
pins (like the rivets we saw in Chapter 1). Examples in the UK
include the Menai and Clifton Suspension Bridges.

Though these chains were, of course, a lot stronger than ichu
grass, they caused design and construction challenges. Each link
needed to be cast or forged, requiring a lot of material and energy
to heat up the iron and liquefy it. The resulting plates were heavy
and tricky to handle. In order to create space for workers to put
the chain together, one link at a time, temporary platforms needed
to be constructed where the final bridge would eventually sit.
The weight of the chains meant there was a limit to how long the
span could be, otherwise the chain would struggle to carry its own
weight, let alone the weight of the bridge suspended from it. There
was also the added worry that the failure of just one of the pins
could lead to the failure of the entire chain: unlike with string,
there was limited redundancy, meaning that one failure could have
catastrophic consequences.

Replacing chains with wires diminished these drawbacks. Once
we figured out how to make steel cheaply during the Industrial
Revolution, wire could be produced in large factories, then transported to cable manufacturers. They used machinery to pull and
stretch the wire to make it thinner without needing to heat up the
steel (a process called cold-drawing). By weight, such cold-drawn
steel wire is a significantly stronger material than its older cousin,
wrought iron, so cables made from this newer material are significantly lighter than iron chains.

Once the wires are reduced in diameter, typically to a few mil-
limeters, they are then bunched into strands, in much the same
way as multiple lengths of yarn are twisted together to create

string. With steel, however, there are more options. The wires can
be twisted together in layers, just like yarn, or they can be laid in
a straight line in parallel and then clamped together. It depends on
what you will eventually be using them for.

To see some cables, I traveled to visit engineers at Macalloy,
a firm based just outside Sheffield, where bars and cables are
adapted for a range of different structures. The solid bars used for
my first project, the Northumbria University bridge, were in fact
supplied by this very factory, but today I was looking at the twisted
variety. Surrounded by clanging metal, puffs of steam, and the
smell of grease, I was taken toward one end of their factory, where
a big shelf was stacked with cables wrapped around spools—rather
like thread around bobbins, but on a much larger scale.

I scrutinized the end of one of the cables to see what it looked
like in profile. There was a clear Z-twist visible from the outside. The cable was made from seven strands of smaller cables,
which were themselves twisted together. Carefully separating one
of these seven strands, I could see that this one was made up of
nineteen wires: a straight one through the center, wrapped by six
wires in a Z-twist to create a hexagon, around which was a layer
of twelve wires in an S-twist, forming another hexagon. Since the
outermost layer was an S-twist, it made sense that these strands
were, in turn, wrapped around each other in a Z-twist to form the
final cable. With fuzzy eyes from figuring out the twists, I took a
breather and found out a bit more about what these twisted cables
are used for.

One of the biggest advantages of twisted cable is its stability
and flexibility. Once the wires are twisted together in whatever the
required configuration, the cable can be curved and moved without
them coming apart. This means they can be wrapped around spools
or drums, which is a conveniently compact and efficient shape for
transportation. It also means that they can be as long as you prac-

tically like (while making sure that the spool doesn't get so heavy that it can't be lifted). Twisted cables are ideal for the smaller end of structures supporting heavy pieces of art from ceilings in galleries, outdoor canopies made from stretched fabric, hangers for staircases, or the glass elements of facades. Smaller bridges use them as well. One of the most unusual applications I learned about at Macalloy was the 400-meter-long zipline that hurtles you through a roller coaster at Ferrari World in Abu Dhabi.

The disadvantage of the twisted cable is that the slight give or movement that allows it to be flexible in the first place means that, when pulled with large forces, it stretches and relaxes a little. The cable is also not as strong as it could be if the same wires were placed in parallel. So, for the world's largest bridges, where the loads from the deck and the vehicles that drive over it are immense, the cables tend to be made from parallel-strand cables.

One of the first bridges in the world to use this technique was the iconic Brooklyn Bridge in New York. (I have a particular fondness for this bridge because, following the death of the original designer, John Roebling, and the increasing ill health of his son, construction continued under the management of Emily Warren Roebling. It's an extraordinary story—particularly at a time when women were viewed as fit only for domestic duties—which I recount in my first book, *Built*.) As well as designing bridges, John Roebling ran a steel-wire manufacturing business, and it was here that the wires for the Brooklyn Bridge were made. He employed around 350 workers, who ran five mills that spun out about three-quarters of the US's wire rope at the time. A reporter who visited wrote: "It was a rare sight to watch these busy workmen taking blocks of red hot steel in their tongs from white heat furnaces, passing them through rolling mills which stretched them until they lay upon the iron floor like interlacing snakes in bizarre shapes, ready to be carried by other hands to the annealing furnaces, and thence through other draw plates until the wire was prepared to bind together either the delicate handiwork of the jeweler or the two cities of New York and Brooklyn with their millions of inhabitants."

The Brooklyn Bridge is a suspension bridge, meaning it has two main cables from which the deck is suspended. To support these two cables are two tall towers at each bank of the river. The cable starts off anchored strongly to the foundation at one end, rises up over the first tower, and then bridges across the river to the second tower. It is anchored again to another foundation behind the second tower. Between the two towers, it sags under its own weight to form a curve called a catenary. To complete the system, smaller vertical cables are hung off the two large cables, and the deck is attached to the vertical cables.

Installing these cables was a feat in itself. They didn't come to the construction site in one piece; rather, they had to be assembled in the air from thinner lengths of wire. First, a temporary steel rope, called the working rope, was pulled over each tower. Using this, a continuous length of a thinner wire was pulled back and forth from the foundations and over the two towers at each end of the bridge. Once this was done 278 times, the 278 wires were tied together to make a so-called strand. Then the process started again. After nineteen strands (each with 278 wires) were complete, these were bundled together to create the final cable. It was wrapped tightly in a skin of soft iron wire for protection. Each cable contained just over 5,656 km of wire.

Since then, the strength of the individual wires has steadily increased, but the design remains similar: wires are bunched and tied into strands in counts of numbers like 127 to form a hexagon, and these strands are also bunched into hexagons. They are tightly clamped together to create that friction. To prevent the wires from rusting, dry air is pumped into the cable at one end to dehumidify the air in the gaps between the wires. Where the cable curves into a dip, any water that collects there is removed. These cables, unlike chains, don't mind at all if a wire breaks here or there: there are dozens, if not hundreds, of other wires that do the job.

Even with the strength increases that we've seen over time—a wire that is just 1 mm in diameter is now strong enough to carry a male gorilla—there is still a limit to how long the cables can be. At some point, the weight of the steel is too much for the cable to carry itself. Engineers are looking at the possibility of using fibers made from carbon. This incredible material is as strong as steel, but much, much lighter. Development is still in the experimental phase, with researchers looking at how to manufacture it effectively. Carbon is not as bendable as steel wire, so at the moment transporting it is difficult. If, or rather, when, we can surmount these challenges, we'll be able to bridge distances that are currently impossible.

Utilitas

It seems intuitive that these structural cables, made of metal and assembled to make the most of their strength, should be strong. But it's less easy to imagine how their nonmetallic counterparts could be strong in a very different way. Such is the flexibility of string that when woven from steel, it can hold up bridges, and when woven from plastics, it can hold off bullets.

The inventor of that material was, like Emily Warren Roebling, a woman working in what was traditionally seen as a man's world. Stephanie Kwolek was the daughter of Polish immigrants to the US. Influenced by her father, who was a naturalist, she studied science with a major in chemistry, with the intention of becoming a doctor. But a medical degree took a long time, making it difficult to cover the costs. Kwolek made a practical decision: with so many men still overseas after World War II, new opportunities for women were opening up in industry. In 1946, she got a job at the chemicals company DuPont and dropped the idea of pursuing medicine, although she ended up saving more lives than she ever could have as a doctor.

DuPont had been conducting research into replacing steel wire in vehicle tires to make them lighter, and Kwolek was assigned the task of creating fibers to test. She had been experimenting with polymers called poly-p-phenylene-terephthalate and polybenzamide. Polymers are very large molecules that are made up of many small units that repeat, just like an iron chain is made up of many identical links. To create the long-chain versions needed to make fibers, the polymers were first dissolved in a solvent. Next, the resulting liquid was spun in a machine called a spinneret, a device not unlike a cotton candy machine, which causes the excess liquid to separate out from the fibers that were mixed within it. Normally, in this kind of experiment, the mixture of polymer and solvent created a clear and viscous liquid, but Kwolek's combination produced a surprisingly watery and cloudy mix. Despite reservations that her sample was faulty, Stephanie decided to test it anyway. Against expectations, what emerged from the milky mixture was a very strong, stiff fiber that, unlike nylon (the first synthetic fiber, invented in the 1930s), didn't snap easily.

DuPont realized the practical significance of this new fiber, polyparaphenylene terephthalamide, and named it Kevlar. Kevlar is very light for its strength. If you measure its strength and divide this by its density, and do the same for steel, Kevlar wins by a multiple of five, which means that, for the same strength, Kevlar is much lighter. It came into commercial use in the 1970s, initially in the tires of racing cars, much as the original research intended. But that combination of strength and lightness meant that Kevlar was a material of supreme utility. Kevlar has since found its way into an extraordinarily wide variety of applications, from sports clothing, tennis racket strings, car brakes, and bridge cables to smartphones, snare drums, and fiber-optic cables. Most of us, however, know of this material thanks to one object—the bulletproof vest. Of course, Kevlar is not the first material to provide defense against weapons; in medieval times, knights clad themselves in sheets of metal, but these took a long time to make and were heavy and difficult to move in. Reliance on some form of metal plates for protection continued through the First and Second World Wars. Flak jackets made of nylon were tested during the Second World War, but they, too, were cumbersome and not particularly effective. Kevlar revolutionized protective clothing, not just because of its extreme strength—even metal bullets cannot penetrate it—but because of its usability: Kevlar's lightness, flexibility, and heat-resistance make it the perfect material for protective use.

Kwolek's discovery and development of Kevlar is all the more noteworthy because it took place in an industry that, at the time, was extremely male-dominated, with limited opportunities for women. I wondered how she found working in such an environment, and found this quote in an interview that summed it up: "I was fortunate that I worked under men who were very much interested in making discoveries and inventions. Because they were so interested in what they were doing, they left me alone. I was able to experiment on my own, and I found this very stimulating. It appealed to the creative person in me."

Of course, a bulletproof jacket is not a Chanel jacket: it's hardly a fashion item (unless you're into motorcycle gear). All the same, it serves as a reminder of how much we have depended on string and other forms of woven material to clothe and protect us, even if just from the elements. It's difficult to establish when exactly humans started wearing clothes made from string, because few samples have survived in the ground. Rather than woven cloth, our ancestors first covered their bodies with animal skins, furs, grass, and leaves. It's likely, though, that some of these would have been stitched together: primitive sewing needles have been dated back to around 40,000 years ago. But evidence that we've probably been wearing clothes far longer than that comes from a surprising and slightly gruesome source: body lice. While head lice live and feed only on the scalp, body lice focus on the rest of our skin, but they live in clothes. Some scientists hypothesized that if they could figure out when the body louse originated, that might indicate when humans began wearing clothes fairly commonly. The answer they came up with was around 72,000 years ago, plus or minus 42,000 years.

Making clothes from natural materials entails five key steps: cultivating and harvesting the plant, preparing the fibers, spinning them to create yarn, weaving the yarn to form textiles, and finally, putting it together to create clothes. The first solid evidence of woven textiles dates to around 6000 BCE. In Çatalhöyük in Anatolia, archaeologists found a baby's skeleton wrapped in a piece of linen. Between 5000 and 3000 BCE, Egyptians and Indians developed the art of spinning linen and cotton to create long, strong thread.

From there, we went on to domesticate sheep and produce wool, and also to manufacture silk. Cloth made from silk, wool, cotton, and hemp dominated the trade along the Silk Road for the nearly 2,000 years it operated. So far, the whole process was carried out manually on a relatively small scale, with skills being passed down from one generation to the next.

Then, the late sixteenth century saw the seeds of industrialization planted with the invention of the stocking frame, a mechanical knitting machine. Development of such machines exploded in the West in the eighteenth century. The flying shuttle improved the looming technique to enable quicker weaving of cloth. The spinning jenny was the first machine to allow thread to be spun quickly and in quantity. Stronger threads for yarn could now be produced on spinning frames, and other inventions enabled greater control over the weaving process, automation of cleaning cotton fibers, and looms that could weave complex designs.

This era saw a huge increase in the output of machine-made textiles in Britain, which had a lasting impact on its colonies. Every engineering invention is heavily influenced by societal power structures, and the woven strings with which we clothe our bodies offers a good example of this. The British East India Company used its colonies, particularly India, as a market for its industrially produced textiles, slapping high tariffs on Indian handmade exports and purchasing raw materials like cotton at exploitatively low prices to drive profit into Britain's economy while crippling its colony's. That's why the spinning wheel—the charkha, which I wrote about in Chapter 2—became an enduring symbol of India's independence movement: it was a way of encouraging the people to reject British products in favor of local product, creating personal financial liberty and protesting nonviolently.

In addition to the enduring impact of string on prosperity, poverty, and power in terms of the economies of countries, it has also affected societal "norms" of gender. As an engineer, I am interested in how these norms influence the choices that children of different genders make in life, and the direct impact these have on biases in the world of work, including the huge gender gap in my profession. I'm fascinated (and often appalled) by how, somewhere along the way, ideas and stereotypes around gender, race, religion, class, and caste became entrenched in the simple fact of what we wear.

Alok Vaid-Menon, a gender-nonconforming artist, performer, poet, and author of *Beyond the Gender Binary*, says: "We should be able to decide what the clothes and colors we adorn, the bodies we inhabit, and the people we love mean to us. There should be no boys' clothes or girls' clothes, just clothes." Vaid-Menon argues,

"fashion should proliferate possibility, not constrain it." But historically, clothes have been used to create and entrench the so-called
differences between men and women. A striking example is the corset, where women in the West had their bodies literally forced into
defined shapes to accentuate femininity. In popular culture, we see
depictions of lacing—the string that is interlaced at the back of a
corset—being violently tightened to achieve the perfect silhouette.
Of course, what "perfection" looked like changed during the nearly
four centuries that they were worn: from creating a high-waisted
profile, to a triangular one, to the Victorian norm of a curved waist
close to the hips. Here, string was utilized to oppress, to reinforce
a set of limiting societal standards.

Given the stark contrast between girls' and boys' clothes I see
in shops—I usually have to go to the boys' section in clothes shops
to buy my daughter outfits that feature excavators and trucks—
it's important to remember that this wasn't always the case: that
"norms" evolve and change. In Hindu mythology going back millennia, women and goddesses were often depicted bare-chested,
and men and women both wore loincloths, kurta-pajamas, and
anghrakhas (long, dress-like robes). Many modern designers in
India are reexploring and embracing this heritage and creating
gender-fluid collections.

In the West, even just a couple of centuries ago, young boys, like
girls, wore long dresses that were not typically split into gendered
colors. In fact, pink was more often used for boys, as a shade of
the military color red, while girls more often wore blue, the color
worn by the Virgin Mary. Pink cloth required expensive imported
dyes and was worn by sixteenth-century men to signify financial
strength and even physical bravery. While there was some fluidity
for young children, there were many laws in the US that controlled
who wore what, like the one that prohibited women from wearing
pants (which was only repealed by the attorney general in 1923).

These have an impact on us even today: dresses and skirts are still deemed the appropriate attire for women at weddings or balls. On the other side of the story, although it has become largely normal for women to wear pants and other clothes once considered male, it is less "acceptable" for men to wear dresses, skirts, or makeup, all of which are still considered mainly female. The pink-blue divide (now swapped around) remains entrenched in children's clothing and toys.

The other big societal impact that our use of string in clothes has is ecological. The greenhouse gas emissions from textile production are more than those of all international flights and maritime shipping combined, and textile manufacturing generates around 92 million tonnes of textile waste and consuming around 1.5 trillion liters of water each year. In addition to looking at our consumer habits, we are also turning to technology to address this massive issue, from the manufacturing process to recycling. In terms of fibers themselves, innovators are using materials like pineapple leaves, apple peels, grape skins and stalks, and wood pulp to create leathers that contain no animal products. Others are using marine plastic and plastic bottles to make yarn that can be used to make cloth. Moving to natural fiber hemp helps, as it needs 50 percent less water to grow than cotton and doesn't require harsh pesticides.

Cloth also plays an important role in our health. During the COVID-19 pandemic, scientists and engineers extensively tested different materials to ascertain the level of protection they could offer against breathing in the virus. Medical or surgical masks are generally made from three layers of plastic-based material that is "nonwoven." Traditional material for clothing or furniture has a woven or knitted structure that creates a regular pattern, but nonwoven materials have a random arrangement of fibers, like spaghetti on a plate. The randomness makes the material much better at catching small particles, like viruses. A type of this material is spunbond polypropylene, where the random fibers are compressed and melted together—this material is often found on the underside of furniture. It's washable and generally wasn't part of the medical equipment supply chain, so, in some places, it became the recommended filter to add between layers of a cloth mask for added protection.

Given the ubiquity and utility of fibers in human life, focusing our attention on them to make our health, lives, and world better makes perfect sense. Meanwhile, the next time I go shopping for clothes, I'll be thinking about more than their price or how they look on me: I'll be considering their role in power structures, their physically restrictive nature, and their environmental impact, but also their potential to free us all from the boxes in which we inadvertently find ourselves.

Venustas

A few years before I stood on my (yes, it belongs to me) footbridge, I experienced another life-defining moment. It was August 1999, nine years after I began training in Bharata Natyam, a classical Indian dance form that has been passed down through generations for over 3,000 years. I was performing my arangetram, a ceremony that marked my maturation as a dancer. Now, over two decades on, I still remember the moment I stood in the wings on stage as I waited for my cue, with over 200 of my friends and family waiting eagerly in the audience. I stood up straight, hands on hips, shoulders back, forcing a smile. A beautiful swirl of sound filled the auditorium as musicians plucked the tanpura and tapped the bronze cymbals. Rapid beats on the dholak signaled my cue. A deep breath, and I took strong, rhythmic steps onto the stage.

The music that was such a defining feature of my arangetram was dominated by string. Alongside the sustained undertone of the four-stringed tanpura and the beats of the rope-tightened dholak, there was a violin played with a strung bow for melody. The two heavy bronze cymbals (connected by a white cord) clashed, complementing the sounds of the ghungroos, the collection of dozens of small bells tied around my ankles.

The flexibility and strength of string in tension, which makes it good for clothing, is also what makes it perfect for music. In violins, sitars, cellos, sarods, and pianos, strings held taut are struck, plucked, or bowed to produce notes. The quality of those notes—their pitch, tone, depth, longevity—are underpinned by the physics of waves.

When you tie a string between two points and tweak it, it vibrates. The vibrating string causes the air around it, and the body of the instrument, to vibrate and create waves. These ripples of air from a vibrating string translate as sound in our eardrums.

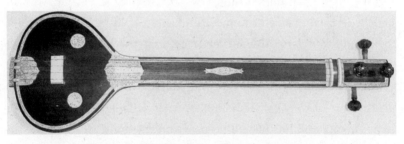

Tanpura. By Allauddin, Creative Commons

I was always more interested in dance, and regretfully never learned to play any instruments, but I knew that a vital part of the ensemble at my arangetram was the tanpura. This instrument comes in varying sizes, typically between 1 and 1.5 meters. The base is its bulbous body, which is made from a dried gourd or pumpkin. A long wooden neck extends up from here, with four pegs at the top. Four long strings are wrapped around these pegs, then drawn down along the neck to the main body. Here, the strings are supported by a wide "bridge," then proceed to wrap around the curve of the pumpkin. Toward the end of each string is an ornate bead, which is in contact with the body of the instrument. This can be adjusted up or down the curve to control the tightness of the string.

The tanpura does not have any frets; it is not used like a guitar or violin to play melodies. Unique to Indian classical music, the tanpura creates an acoustic soundscape, a continuous base layer of tone that inspires the rest of the musicians to play. I had always wondered how it makes that remarkable sound and decided to find out more. It turns out that it comes not just from the musician's dexterity, but also from a careful and unique use of string.

On what happened to be the day of Diwali, the Hindu festival of light, I traveled to Southall in London. It's an area known for its large Punjabi population, and I was excited to see streets full of people dressed in salwar kameez and saris, holding plates of street food and sweets, chattering away against the backdrop of shops with loudspeakers advertising their fireworks.

After picking up a plate of pani puri, I made my way to my destination. JAS Musicals is a shop that sells and repairs Indian instruments and has been a presence on the main street in South-all for decades. Inside, I found Harjit Shah, its owner. He wasn't much taller than me, and wore a dark blue turban framing a dense, graying beard. Around him were shelves filled with instruments—sitars, veenas, tablas, harmoniums, tanpuras. We clattered down some narrow stairs into his workshop in the crowded basement, where we sat and Harjit told me his story.

He originally trained in India as an engineer, but in 1984 he came to the UK with his father. It was a trip that ended up turning his life upside down because, back in India, the prime minister,

Indira Gandhi, was assassinated, which meant they weren't able
to travel back home safely. His family encouraged them to stay in
London and set up a life there rather than return to a precarious
political situation.

Harjit did whatever he could to earn money, including deliv-
ering milk and driving taxis. One day, the gurdwara where his
father was a priest asked him to get four harmoniums from India.
He ordered them, but because they were made for a dry and hot climate, the keys seized up in London's foggy cold, which gave Harjit
the idea of setting up a business selling Indian instruments. Starting out in the garage of his rented home, he put his engineering
skills to work, making modifications to the instruments so they
would work well in a European climate.

"When I listen to music, I listen to the instrument, not the
melody. I'm taking in its acoustic properties. Somebody might be
playing incredible music, but my ears, my body, my brain enter
another world. I'm focusing on the instrument: how the musician
is plucking, how the strings are reacting, how often they're tuning, how stable the tuning is. Are they steel strings, brass strings,
gut strings?"

Back upstairs in his shop, Harjit climbs a ladder to bring down
a large tanpura from the top shelf. It has three steel strings and
one brass, which plays the lowest notes. Placing the instrument
horizontally on a table, he begins to pluck the strings one by one,
tightening the pegs at the top in sequence. I can hear the pitch of
the sounds changing. Experience tells him when he has achieved a
reasonable level of tightness in the strings. But this is just step one.

Harjit continues to pluck the strings while gently shifting the
beads at the bottom of the strings to fine-tune, back and forth. So
far, all this looks familiar, even to this non-musician. Once satis-
fied, he turns to me and says, "Now I'll show you the real magic of
the tanpura."

Harjit takes out a spool of cotton thread, the type you'd mount on a sewing machine, and breaks off four pieces about the length of his forefinger. He plucks the first steel string. I hear the sound of his finger leaving the string, then a note, which quickly dissipates to silence. "An accurate but flat note with no life," he says.

He takes a piece of the cotton thread and places its middle below the steel string. He lifts up the ends of the thread and holds it between his forefinger and thumb, so it is looped around the steel. He drags it slowly down to the bridge. Then, he plucks the steel string while slowly pulling the cotton thread, now sandwiched between the bridge and the steel. Suddenly, the flat notes give way to a buzzing sound. He does this for all four strings and then plucks them sequentially.

Now the buzz of each string continues long after he has plucked it. The notes overlap: it is no longer just one sound but many— varied, rich, otherworldly. It's even spiritual. I can't help but close my eyes; all the sounds from the street outside seem to fade away. I'm transported to the day when I did my arangetram. I can feel vibrations in my temples, traveling down my arms, my back, my legs. I begin to understand how it inspires musicians, and how it gave me the focus I needed to complete my performance decades ago. I'm completely immersed in the sound of the tanpura, this luxurious, shimmering sound that has been magicked out of the tweaking of a piece of thread.

Variations of the tanpura are found in Eastern Europe, Turkey, throughout North Africa, the Middle East, and India, so its history is tricky to track. There are theories that it developed from indigenous instruments in India, or that, centuries ago, Arab-Persian musicians introduced similar instruments. Harjit tells me that in the Samveda—part of the scriptures of Hinduism that date back over 3,000 years—there are passages that can be interpreted to imply that the gods and goddesses themselves developed these forms of instruments based on what they experienced during their meditations. But he quickly backtracks: "Let's not go too deep into spirituality; let's stick to the engineering."

Harjit laments the fact that instrument-makers in India are undervalued, experiencing poor living standards and education, and often ending up illiterate. He is concerned about the loss of skills, and also about the lack of science applied to the development of the tanpura. For an instrument that he believes took many generations to perfect, it seems development has stalled. "There is research on tuning the tanpura, the tones of the tanpura, the voices of the tanpura," he says, "but little on the physics or engineering of the tanpura." With geeky glee at the opportunity to speak to another engineer about the instrument, he goes on to explain to me how the tanpura is able to create these overlapping harmonics.

The key lies in the bridge, the solid piece on which the metal strings rest at one end. In Western instruments, the bridge tends to be quite narrow and flat, so that the strings come clear off the edge of a sharp corner in order to narrow the pitch of the note as much as possible. The tanpura's bridge, on the other hand, is wide

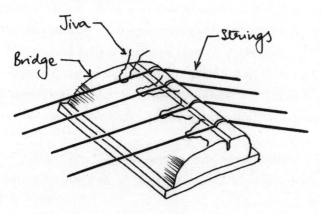

The curved bridge of a tanpura with jivā strings

and its top surface is curved. Musicians and makers of the tanpura refer to javārī, which describes the quality of the instrument's tone from an acoustic perspective. It is also a verb, used to describe the process by which the maker shapes the rounded bridge to produce the desired sound. Originally made from ivory, tanpura bridges are now often made from ebony or a form of nylon.

The curved bridge does something interesting to the string. When a string is plucked, it starts to vibrate, rapidly going up and down, while gently grazing the surface of the bridge. The physics of waves tells us that longer strings produce sounds with longer wavelengths, which means a lower pitch. Likewise, the shorter the string, the higher the pitch. In Western instruments that have sharp bridges, the end point of the string—and so its length—is always the same. However, in the tanpura, the point of connection between string and bridge varies depending on its motion. When the string moves up, the length of the string is very slightly longer than when it vibrates down. Also, as the energy of the string diminishes, the point at which the string contacts the bridge slowly changes. These subtle changes in the length of the string start to produce overlapping notes.

The cotton threads add another dimension. When Harjit moved the thread slowly between the string and the bridge, he arrived at a point where the string begins to resonate or vibrate dramatically, releasing a cascade of notes. The cotton thread pinpoints that very specific position on the bridge that allows the metal strings to dance freely, giving them, and the sound they make, life. Harjit told me that these threads are known as jivā, which translates as life or soul.

Cotton thread is not the only woven material that has played a surprising role in the music-making of stringed instruments. The strings themselves might be made not from steel and brass (like the tanpura), or from silk or nylon, but from a far grislier substance—catgut. Although this is not actually from cats, it is indeed made from guts. Typically, a whole lamb's intestine is collected from an abattoir while still warm in order to harvest its collagen, a fibrous protein found throughout the bodies of mammals, particularly where strength and elasticity are needed, such as in our skin, cartilage, ligaments, and tendons. The lamb's guts are soaked in a series of chemicals to dissolve away all the tissue except the collagen fibers. Once cleaned, they are stretched, twisted, and dried under tension. Although catgut strings can be fragile and break down more quickly than steel or synthetic string, they are preferred by many professional musicians. The combination of the mass of the catgut and its flexibility produces an arguably richer sound. (Catgut is valued in other disciplines, too. Some professional tennis players choose catgut for their rackets over polyester or Kevlar because of its combination of strength and give. And the sutures used to stitch us up post-surgery were once made from gut because, unlike thread made from silk or nylon, they dissolve into your body over a few weeks.)

Whatever material is used for the strings on a musical instrument, the way in which it's constructed is often more complex than might at first be imagined. If you touch the strings of the tanpura or the top three strings on a guitar, you'll see that they're effectively plain string: long lengths of thin metal or (for classical guitars) nylon, precisely made so their thickness is consistent throughout. Touch the bottom three strings on a guitar, however, and you'll see that they're wound, to give them more mass and lower pitch. A wound string has a plain string at its core, with one string (or more) wrapped around it in a helix shape. The strings with higher pitches are generally wound less than those with lower pitches. Today, the process of making string is highly automated, meaning that manufacturers can churn out over 7,000 strings daily. I think it's important to remember that behind all this technical detail and manufacturing engineering is the quest for beauty: to create the perfect sound to touch our senses.

Even in this quest for beauty within musical instruments, the ideas of firmitas and utilitas are essential, too. To produce that note, you need to tune the string; you tighten it, turning a screw to add more and more tension into the body of the fibers until you're fearful that it might snap. But it doesn't. The string resists these forces with strength to produce the useful output we're after, without which music doesn't exist. With this extraordinary intertwining of flexible strength with practicality and aesthetic, string has become a defining factor in our culture. Far from being an innovation that is known for simple pragmatism and usefulness, string shows us how engineering can bring these principles together with beauty, to create truly marvelous experiences that have certainly made my life, at least, richer.

Pump

Five millennia ago, in ancient Mesopotamia (in what is modern-day Iraq), the landscape consisted of barren plains, dotted with areas of marsh, that extended between the rivers Tigris and Euphrates. The arid conditions meant that early settlers needed to irrigate crops in order to feed their growing population. Invention often comes as a response to need, and for the ancient Mesopotamians, their plight was one that humans have come up against for time immemorial. In a stroke of genius, an unnamed inventor—or perhaps a team of inventors—came up with something that would enable them to grow crops in abundance and feed all. This invention was a crane-like structure called a shadoof. Like a seesaw with a very tall support, the shadoof is an upright frame with a lever on top: a long pole with a bucket attached to one end and a counterweight at the other. This simple and ingenious structure allowed them to raise water up in the bucket from the rivers and distribute it to land—an act vital to their survival.

The shadoof is a pump. The word pump may conjure an image of a fairly sophisticated piece of technology requiring many moving parts,

but, at its most fundamental, a pump is a contraption that moves liquids or gases. Pumps can be as straightforward as a bucket of water
at the end of a string that you pull, or as complex as a multi-piston,
motor-driven engine that moves and burns gasoline to power a vehicle.

Historically, pumps have played a vital role in bringing clean
water to us, taking dirty water away, and enabling us to grow food
in bulk and in hostile environments. The reason we need them is
because fluids (this includes both liquids and gases) naturally move
in a way that responds to the forces around them. They flow from
higher to lower levels, like a waterfall does, due to gravity. They
also move from high-pressure to low-pressure areas, because free-
flowing particles don't like inconsistency; they want to be in equilibrium. (The air inside a balloon doesn't like being compressed
relative to the air outside it; that's why, if you don't tie up a balloon
after inflating it, the air rushes out to create balance.)

Pumps force fluids to act in unnatural ways. This could be pushing them up against gravity, forcing them into a high-pressure situation, or even just transporting them somewhere else. Like the
spring, the pump comes in many different forms. They have been
developed independently all over the world in response to different
contexts and needs: a testament to the long and geographically
diverse history of engineers exploring the pump's potential as a
solution to engineering problems.

Innovative pumps like the shadoof, and the Archimedes screw
(which was invented in ancient Egypt and discovered by the
Greeks—mentioned in Chapter 1), originally arose in arid regions
where people had to be creative about their water supply. Even in
the modern world, those of us lucky enough to live with clean water
that pours freely out of our taps are reliant on pumps. I find it fascinating (and also a shame) that many of the pumps we use today
trace their origins to the Middle Ages: more specifically, to an engineer whose name may not be familiar to us.

In his treatise *Kitāb ma'rifat al-hiyal al-handasiya* (*The Book of Knowledge of Ingenious Mechanical Devices*), published in 1206, the inventor and engineer Badi' az-Zaman Abu-'l-'Izz Ibn Isma'il Ibn al-Razaz al-Jazari described in minute detail more than fifty mechanical devices. Based in Diyar-Bakir (in modern-day Turkey), where he served the Artuqid kings for several decades as a mechanical engineer, he was known as a master engineer and craftsman.

Al-Jazari's encyclopedic treatise includes six main categories of machines and devices. Not only does he describe and illustrate in detail how they work, but he also meticulously explains how to construct and assemble them, providing future engineers with a treasury of knowledge. Bringing together form and function in a spectacular way, one of his best-known constructions is the ornate and colossal Elephant Clock (which is four feet long and six feet high). It incorporates a plethora of his creations—systems like automata, flow regulators, and closed-loop systems—that are still used in engineering today.

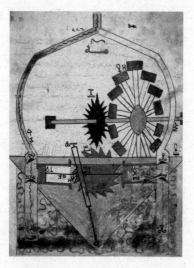

One of Al-Jazari's pump designs

Al-Jazari also invented pumps to raise water in the dry climate of his homeland. One particularly clever version was what is known as a reciprocating pump, a pump that fulfills two pumping actions at the same time. He arranged two copper cylinders opposite each other. Each had plungers that were connected by a single rod. A gear attached to a swinging arm pushed the rod back and forth, so that in one stroke, it pushed water through one cylinder, while pulling water into the other. This was believed to be the first true use of a suction pipe, where liquid is pulled into a partial vacuum.

The crankshaft, a machine that converts rotational motion into linear motion, is a central feature of fuel-powered vehicles today, and is attributed to al-Jazari. The device has a main arm with a series of straight rods coming off it perpendicularly. When the main arm is rotated, because of its "cranked" shape, it pulls the rods back and forth in a line compressing gas or fuel. This technology was central to the invention of the steam engine and car engine; without it, our world wouldn't look the same.

Our need for and dependence on water has meant that a lot of pump development has been focused on moving it. In around the seventeenth century, engineers turned their minds to other liquids and other uses for pumps. As industrialization increased in the nineteenth and twentieth centuries, pump design developed rapidly, and pumps were used in a broad range of applications. Cylinders with plungers—in other words, pistons—such as those used by al-Jazari were applied, for example, to bicycle pumps and car engines. Rotary pumps with a spinning component moved the fluid along, like the gear pump (which pushes viscous liquids like gasoline). Pumps became more sophisticated and complex as manual operation gave way to electricity. Centrifugal pumps used this new source of power to spin liquids around at high speeds, forcing them outward by means of centripetal force (in much the same way as

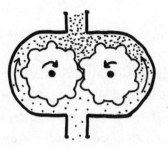

Gear pump

people are pressed against the cylindrical walls of spinning rides sometimes seen at amusement parks), and then releasing them into an outlet.

While there are endless examples of pumps around us that have made modern life possible, one of the most sophisticated pumps existed in nature long before we invented our own versions. Within each of us sits an extraordinary pump, without which we cannot live. The heart is the first organ to develop in a fetus, because the fetus's survival depends on oxygen and nutrients being transported to all cells and on waste being taken away. Once we enter the world, the heart and the circulatory system does this, day in, day out. Our hearts beat around 100,000 times a day to ensure that our bodies can function. Even though engineers have invented and built ingenious pumps throughout history, they have long struggled to create pumps that work with a similar robustness and efficiency, just as they have long sought ways to repair the heart when it begins to deteriorate or fail. The problem is that the heart is not a classical pump with standardized moving parts; it is not a device with dimensions rigidly laid out in an engineering drawing. We've been able to replicate it, up to a point, and save many lives. But it took a lot of experimentation and failure—and dead cats— before we could do this.

What makes the heart such an extraordinary pump is its versatility, reliability, and longevity. It is only the size of a fist. When everything is running smoothly, it pushes 5 liters of blood a minute. It can very quickly and automatically increase this to around 20 liters when needed: if we're running, scared, or in any other situation where the body needs a boost of oxygen. Elite athletes can have rates of almost double that. The heart does this not only by increasing the frequency of its beats but also by adjusting its size. It operates relentlessly and reliably for the length of our lives. Over an eighty-year lifetime, it pulses 3 billion times. And our hearts have done this, at least historically, without ever being stopped for maintenance.

The heart is a muscle, driven by electrical signals generated by our nervous system. A healthy heart has four chambers, which are hollow: two atria at the top, and two ventricles at the bottom. The right side of the heart receives blood that is depleted of oxygen from the body and sends it to the lungs, where it picks up fresh oxygen that we've breathed in and expels carbon dioxide. This freshly oxygenated blood then pours into the left side of the heart, first to the atrium, and then down into the left ventricle. This part of the heart is the strongest, since it has to push the blood all the way to the tips of our fingers and toes.

But what happens when our personal pump goes wrong? Over coffee, cardiologist Rohin Francis explained to me how vital the heart and circulatory system is in keeping us going: if the brain is starved of oxygen for just a few minutes, irreversible damage begins to set in. After around ten, it's curtains.

One of Rohin's enduring memories in his career so far was standing next to a woman whose chest was open—and empty. Her heart and lungs had been removed, and Rohin could see the back of her ribcage through a gaping hole. This woman was alive. She was having a transplant of her heart and lungs, or cardiovascular system, because a heart defect she had been born with nearly sixty years earlier was finally making her very ill. There was a twenty-minute delay in the donor organs arriving from another hospital, so the lead surgeons and many of the team had left the operating room for a short respite. Rohin stood there, absorbing the enormity of the situation, wondering how many times in history someone had seen this sight: a human without their heart and lungs, alive. His thoughts were punctuated by the whirring of a large machine behind him that was doing the job of her missing organs.

As one of the most complex organs of the human body, the heart can come with an abundance of defects. Like this woman, thousands of babies are born every year with a hole in the heart, a defect in the wall that separates the two sides of the heart. In many cases, no intervention is needed, but in others, the hole needs to be closed soon after birth. For years, this was almost impossible to fix surgically; surgeons were severely time-restrained if attempting procedures while the heart was stopped: even if the patient's body was cooled down to reduce how much oxygen the brain needed, they only had about ten minutes to get the necessary work done. The only other option was to perform surgery while the heart was still full of blood and beating. Theoretically, a hole in the heart could easily be fixed with stitches or a graft, but practically, any attempts made were almost futile. At best, surgery made little difference; at worst, patients didn't survive. And so, the search was on for ways of keeping the brain and other delicate organs alive while the heart was unable to do so.

The machine in the operating room keeping Rohin's patient alive was the extraordinary heart-lung machine, which to most medical professionals is known simply as "the pump." In February 1930, a young medic, John Heysham Gibbon Jr. (more usually known as Jack), became a fellow in surgery at Harvard Medical

School. Despite not having much experience in experimental surgery research, he was set to work in a small laboratory. Fortunately, he was assisted by an excellent and experienced technician,
Mary (known as Maly) Hopkinson, who initiated Jack into experimental surgery and was in no small way responsible for the successful projects that ensued.

In October that year, a patient who was undergoing what should
have been a routine and uneventful surgery developed alarming
symptoms of an unusual complication: the lodging of a large blood
clot in the artery that delivers deoxygenated blood into the lungs.
Overnight, Jack watched helplessly as the patient struggled for
life, her blood becoming darker from the lack of oxygen. Like many
patients with heart complications before her, there was no help to
be rendered. He wished there was a way of removing some of that
blood from her veins, putting oxygen in and allowing the carbon
dioxide to escape, and then injecting the refreshed blood back into
her body. In other words, a way of bridging the blockage and performing a part of the heart's function outside her body.

Jack took this seed of an idea and expanded it, aspiring to create
a machine that performed the function of not only the heart but
also the lungs. This would make it possible to operate on the heart
itself, even inside its open chambers, without catastrophic impact
on the other organs. (Unbeknownst to Jack, the Russian scientist
Sergei Sergeyevitch Brukhonenko had also been working on this
problem since the 1920s. He succeeded in bypassing the heart and
lungs of a dog for two hours before unexpected bleeding ended the
experiment, and the dog's life. His research was eventually halted
because of the war. This is another example of a need driving multiple instances of something being invented.)

The two main components that Jack and Maly needed to create were, first, a form of artificial lung to provide oxygen to the
blood and remove carbon dioxide, and second, a pump to propel

the blood through the machine and the body. Designing the pump was a huge engineering challenge. It needed to be efficient, robust, and reliable, and to include backup systems if there were leaks or the power failed. The operator of the pump needed to be able to adjust the flow of the blood depending on what the patient needed. The blood would need to be warmed before it was sent back into the patient's body, as it would have cooled down while circulating through the machine. As if that weren't enough of a challenge, red blood cells are incredibly fragile, as they are essentially tiny bags of fluid enclosed in a delicate membrane. Turbulent or rough flow of the blood could burst them, jeopardizing the whole endeavor. So, while the pump has to be powerful enough to propel blood to every extremity of the body, it mustn't damage the blood. The human heart has had millennia to perfect this delicate tightrope walk. Matching it would prove a challenge.

Having constructed some initial designs for each of the components, Jack decided to begin his experiments on cats. The reasoning behind this choice was that their smaller bodies had less blood to oxygenate. Local authorities in Philadelphia were killing 30,000 strays a year, so he went out at night with some tuna and a sack and returned to the laboratory with unsuspecting subjects. This morbid work wasn't for the faint of heart.

Maly would begin her work early in the morning, spending several hours preparing the equipment for each day's experiment. When all was sterilized and assembled, she would anesthetize the cat and connect it to an artificial respirator so it could breathe. The next step was to open its chest to reveal its heart. The couple—romance in the lab had, by this time, led to marriage—then inserted tubes into the two main vessels coming in and out of the heart to send the blood through the heart-lung machine. They also had to make sure that the blood didn't clot in this unnatural environment, so they injected a compound: a new medical marvel called heparin. They blocked the pulmonary artery (the one that supplies blood to the lungs), switched on the machine, and watched.

With many failed attempts, frustration, and adjustments to the equipment, they finally succeeded in keeping a cat alive for four hours in 1935. In 1939, they announced that four cats kept alive by the machine for up to twenty minutes had made full recoveries. (One of them went on to have a litter of healthy kittens a few months later.) After years of work, they had managed to double the time one could safely stop the beating heart of an animal, and while twenty minutes may not seem like a lot, it is enough to perform the kind of heart surgery necessary to rectify at least some common heart defects.

In 1952, they were finally ready to try it out on humans. A couple of patients, including an infant, tragically died, but it should be noted that they were already very sick, so their deaths may not necessarily have been caused by Maly and Jack's machine. Then, in May 1953, eighteen-year-old Cecelia Bavolek, who had been born with a heart defect, became the first to successfully have surgery while on the bypass machine. She was connected to it for forty-five minutes in total, and it took over her heart and lung function for twenty-six minutes. Cecelia recovered quickly and was rapidly able to achieve a normal level of exertion. She remained completely well. Jack and Maly Gibbon had succeeded, and changed cardiac surgery forever.

When the Gibbons were designing their machine, Jack had at first tried to emulate the pumping action of the heart itself by creating a collapsible chamber with intake and output valves for the blood. The chamber got filled when relaxed and emptied when compressed, but it was very difficult to clean and sterilize all the moving parts and the valves. Contemporary medical understanding, however, suggested that the body could function fairly well if the blood flowed around it continuously rather than in pulses, particularly for a short period of time. Once they were freed from the restriction of having to strictly emulate the heart, the Gibbons moved on to designing and creating a simpler machine, with far fewer moving parts, which was consequentially far easier to clean and maintain.

The design that Jack eventually settled on was a roller pump, which is the same kind of pump used in bypass machines to this day. In the roller pump, the blood from the patient is collected into a clear plastic tube. A portion of the tube is held in a semicircle, within which sits a rotating motor. The motor has an arm, with a cylinder at each end that is free to rotate about its own axle. As the arm rotates, the cylinders massage the tubing. As one cylinder rolls off the semicircular portion of tube, the other cylinder rolls onto it. This action creates a suction in the tube on one side that pulls blood in, and then a propelling action to push blood out of the other end.

Thanks to the heart-lung machine, surgeons can now perform a broad range of procedures on the heart. Depending on the age and health of the patient and the location of the problem, they can repair the valves and the walls that separate the chambers of the heart, as well as defects that babies are born with. They can remove blood clots from the lungs and repair arteries that have bulges that might burst. And, as Rohin had experienced firsthand, the heart-lung machine opened up the world of heart transplants.

But there is a problem: a dire shortage of donors. Rohin explained that the number of transplants being done now was similar to that in the eighties and nineties, despite an ever-increasing population with heart conditions. People are living longer and dying older, so the number of healthy hearts available for a patient to receive has stagnated, which leads to difficult decisions about who gets the limited number of hearts available. Rohin said that tall people often end up struggling to get a donor because you can put a big heart into a small person, but not a small heart into a tall person: it wouldn't be strong enough to get the blood to the longer extremities. (He added that this, and sitting comfortably in economy seats on a plane, are the only two wins for short people.) So, engineers have been working to create an artificial device that can serve patients who are either waiting for a transplant or not eligible for one—in other words, a replacement heart.

The broad range of approaches that engineers have taken to solve the problem of a broken heart is a point of fascination to Rohin. He told me that Paul Winchell, an American ventriloquist who won a Grammy for voicing Tigger in the 1974 film *Winnie the Pooh and Tigger Too,* filed the first patent in 1956 for a total artificial heart—that is, one designed to completely replace the heart rather than augment it—and it was granted in 1963. A few years later in 1969, Haskell Karp became the first person in the world to receive an artificial heart. The temporary device kept him alive for three days while a transplant was sought. Sadly, he died shortly after receiving a new human heart because of infection, but the temporary heart had done its job. It had maintained life in the absence of a human heart.

The device was known as the Liotta-Cooley total artificial heart, named after its designer, Dr. Domingo Liotta, and the surgeon who implanted it, Dr. Denton Cooley. It was made from materials that weren't affected by the body, including a fiber called Dacron, a type of plastic registered to chemicals company DuPont, where Stephanie Kwolek, the inventor of Kevlar, once worked (see Chapter 6). It had two pump chambers to replicate the ventricles, with two passages for blood to enter the heart (replacing the atria), and valves to control the flow of blood through it. It was driven by air, which meant air ducts about as thick as my thumb had to pass from inside the body to the outside, where a pump powered by electricity kept the artificial heart beating.

The Liotta-Cooley heart was only meant to be a temporary solution while waiting for a donor, but on December 2, 1982, a team led by Dr. William C. DeVries attempted a permanent version by implanting into patient Barney Clark a device called the Jarvik-7. (This was named after Robert Jarvik, who, while a student at the University of Utah, introduced three valuable modifications to the design of the pump, which was, at the time, being tested on calves. These modifications were: improvement to the shape in order to better fit inside the human chest; use of a more blood-compatible material; and a fabrication method that made the inside of the ventricles smooth and seamless, in order to reduce the risk of clots.) When Clark woke up after the operation, he asked for a cup of water, then turned to his wife and said, "I want to tell you that even though I have an artificial heart, I still love you." He lived for 112 days with this heart, but suffered with complications, not least the discomfort of being pumped with air by an external machine. However, two of the other patients who received the Jarvik-7, William Schroeder and Murray P. Hayden, lived for 620 and 488 days, respectively, which showed that such a device might provide a long-term solution, although at this stage patients are still vulnerable to strokes and other complications.

Total artificial hearts, or TAH as they are known, have continued to pose a challenge to design. Any hard edges in the device can cause damage to the surrounding tissues, as well as blood clots and infections. They need power or air to feed the device from an external source, so the patient ends up with tubes coming out of their body, which also risks infection. So far, a TAH requires a bulky driver to power it, which needs to be dragged around, although once the patient is healthier, a smaller version can be carried on the back. Since the first implant in 1969, around thirteen different designs are in development around the world. One of these is the SynCardia temporary TAH, developed in Tucson, Arizona. It has been implanted in around 1,800 patients. Just like the human heart, this device has two ventricles and four heart valves, and it is a pulsatile device, meaning that pulses of air that go in and out through pipes pump blood in and out of the ventricles to replicate the heartbeat.

While surgeons and engineers have continued in their efforts to improve total artificial hearts, they have also brought focus to mechanical pumps that would assist the human heart, rather than replace it. A large proportion of the defects in the heart that create the need for transplant lie in the left ventricle, the chamber that pumps oxygen-rich blood to the whole body, so not all patients need a full artificial heart. According to Rohin, for decades scientists didn't fully appreciate the complexity of the way in which the left ventricle contracts. He explained it to me by making a comparison to a famous skyscraper, which excited me: finally, he was speaking my language. In the City of London there is a bullet-shaped tower that is fondly called the Gherkin. It has a distinct aesthetic: from the outside, you can see horizontal rings of structure that form the floors, vertical structures that form the columns, and a pattern of diagonal lines that wrap around the tower to form giant diamonds. (These, in fact, create the stability system for the tower, the structure that keeps the tower stable against wind forces. It's a creative construction that I explore in *Built*.) Rohin compared the layers of muscles that drive the ventricle to these three structural components: there is a horizontal ring of muscles that contract to squeeze it inward, and vertical muscles that shorten it, but in addition, there are also fibers that wrap around diagonally to create a twisting action. These three mechanisms work together to pump blood quickly and efficiently throughout the body, and make the heart's action even trickier to replicate artificially.

Armed with this information, engineers have been honing what are known as ventricular assist devices, or VADs. Most VADs are attached to the left ventricle at one end and to the aorta (which supplies blood to the whole body) at the other. Blood flows from the ventricle into the VAD, and this pumps the blood into the aorta, so it flows into the rest of the body. Since this is a continuous-flow device (meaning it doesn't need air to be supplied in pulses, like the total artificial heart), there is just one thin electrical cable that comes out of the body to a battery pack.

Rohin's first exposure to a VAD as a fresh medical graduate caused him some bewilderment. A senior consultant had asked him to check a patient's pulse. Easy enough, Rohin thought, but when he placed a finger on the patient's wrist, to his embarrassment, he simply couldn't detect that gentle throb of blood indicating the heart was beating away, and he began to doubt his competence as a doctor. Turns out, it was a trick question; a VAD, like the heart-lung machine, circulates blood in a continuous loop instead of in bursts like in a normal heart or in TAHs. So, the patient had no pulse.

The most successful VADs today use magnetic levitation (maglev) pumps. Normally, in a pump that has a rotor (a fan-like device) that spins to move fluids along, that rotor has an axle, supported on a motor that makes it spin. Growing up in India, we had ceiling fans just like this: the blades were at the end of a long cylinder that was forced around when we switched the power on. But having various moving parts that touch isn't very gentle on blood, and damages its cells and structure. Instead, the maglev pump has a levitating rotor that doesn't touch the motor that runs it, in much the same way as a maglev train levitates above its track.

The maglev pump relies on a pinch of electromagnetic magic, and is yet another beautiful example of how combining different inventions can create something quite extraordinary. It has one motor and one rotor. The motor is a hollow cylinder that houses a series of permanent magnets with coils of wire wrapped around them. The rotor is a flat metallic cylinder with blades, which is nestled inside the cylindrical motor. When the motor is powered up, electromagnetic forces from the magnet and electric current push upward on the rotor so that it ends up suspended. The suspended rotor creates a consistent gap all the way around it, through which blood can flow without getting crushed. The magnets within the motor also force the rotor blades to spin to direct the flow of blood away from the ventricle. To make sure the gap remains perfect, the motor sends out thousands of signals every second. Any discrepancy leads to the electrical current—and therefore the electromagnetic force—being altered, so the rotor's position is adjusted. This means that, even when the wearer is moving around, running, or lying down, the magnets make micro-adjustments to their strength to make sure the blood cells never get crushed.

For now, ventricular assist devices are powered by batteries that are external to the body. As the size of batteries gets smaller, designers are working on models that sit fully inside the chest and can be charged wirelessly, like some of our phones. I asked Rohin about the longevity of these devices compared to transplants, and it surprised me to hear that we are starting to get comparable lifespans. On average, transplants last around fourteen years, though one of his patients has had his for thirty-four years. The newer-generation VADs haven't even been around that long, but older models are pushing past the decade mark.

Pump technology has been used not only to extend and alter the course of life itself but also to break the bounds of exploration—allowing us to boldly go where no human has gone before. But during the first spacewalk, even though a pump was there to preserve the cosmonaut's life, it was nearly his undoing.

After twelve minutes floating in space, Alexei Leonov realized there was a problem. His suit pumped air into his helmet for him to breathe, and around his whole body to create pressure. However, when he exited *Voskhod 2* and entered the vacuum of space, his

suit expanded, becoming deformed and very stiff, like an overin-
flated balloon. (No human had entered free space before, so no one
had known for certain how the suit would perform.) His feet had
pulled away from the in-built boots, and his fingers away from the
gloves. Now it was time to reenter the craft, sliding feetfirst into a
space no bigger than a phone booth. But since his limbs had lost all
connection with his suit, there was no way Alexei could grab on to
the tether connecting him to the spacecraft—or anything else. He
had forty minutes left before his life-support systems failed.

Without informing mission control, Alexei effortfully released
about half the air from his suit by opening a manual valve that
kept the oxygen sealed inside. The immense physical exertion in
every small movement was using up Alexei's limited energy, and
he felt his temperature rising dangerously high, going in waves
from his feet into his legs and arms. But it worked. He managed
to deflate the suit enough to gain some flexibility, and was able to
pull himself inside headfirst. At this point, Alexei was drenched in
sweat that filled his boots up to his knees. He was suffering from
dehydration and exhaustion. He had lost about 6 kg in the space of
just half an hour. Nevertheless, on March 18, 1965, Alexei Leonov
went down in history as the first human to walk in space. Four
years later, Neil Armstrong would become the first man to step on
the Moon and make his stirring "One small step" speech. Alexei
Leonov's report to headquarters about his experience was slightly
more matter-of-fact: "Provided with a special suit, man can survive and work in open space. Thank you for your attention."

The pump is an integral and essential part of the suits that
enable us to survive in the extreme environment of space. There
are two main types of space suit: one is worn inside a spacecraft
during launch and reentry, the other is used for spacewalks, or
Extravehicular Activity (EVA) as NASA calls it, with a Leonov-

like undramatic practicality. The suit used for EVA is called the Extravehicular Mobility Unit, or EMU. In both types of suit, the pumping of oxygen is crucial so astronauts can breathe. But the EMU needs to do much more, because our bodies are designed to live under the weight and pressure of the Earth's atmosphere. Catapult us into space, and a very different set of physical rules apply.

As we all know, space is a vacuum. If you were exposed to this absence of air, the liquid that makes up your body would start to evolve into gas, making your body swell and cool down rapidly. Your lungs would be sucked clean of air, and you would suffocate. But that's not the only assault that would take place. The prolonged exposure to unadulterated radiation could cook your organs, while the dust and debris that move through space at extraordinary speeds might pierce you like bullets. Meanwhile, a spacewalk can buffet the body with temperatures ranging from minus 150°C to over 120°C in the sunlight. So, to keep the human body alive, an EMU suit is effectively a mini-spacecraft that contains everything we need to live and work in this harsh setting. It must have the ability to mirror our movement so that we can move and conduct repairs and experiments, create proper pressure, and provide us with water to drink and oxygen to breathe.

A typical suit has three main parts. The outermost layer protects from the changing temperature and micrometeorites. Below that is the restraint layer, which is stitched together from panels of material (similar to our clothes). This gives the suit structure, stopping it from ballooning in the vacuum. The inner layer is the bladder. Made from nylon and coated with a type of plastic, it is impermeable: it prevents any air or moisture traveling from the astronaut's body outward. Oxygen is pumped into the bladder and helmet from a pressurized chamber at the back of the helmet. It travels over the face to remove exhaled carbon dioxide, then over the body to the extremities, picking up moisture from sweat along the way. This pump makes sure that the astronaut can breathe, and that air pressure is distributed equally over the entire body.

For me, though, the more surprising application of the pump is for another purpose. Alexei Leonov literally sweated into his boots because in order to protect his body from the radiation, extreme temperature changes, and micrometeorites, his suit was made from multiple layers of material. NASA was keen to avoid this issue while designing EMUs for the Apollo Lunar mission—but they had other problems. The suits being designed at the time were stiff and bulky, and severely restricted the astronauts' movements. In 1967, the industrial division of Playtex—a company that specialized in making girdles and bras—used their experience to create a spacesuit made almost entirely from fabric. They put one of their employees in their prototype and then filmed him running and kicking and throwing a football in the field of a local high school. They won the contract, giving their bra-sewing seamstresses a new project to work on. (These women were thanked personally by the astronauts who wore the suits they made. In an interview to mark the fiftieth anniversary of the Moon landing, Lillie Elliot, who cut out the patterns, recalled that her heart was in her throat when the men started down the ladder. While most people were in awe of humankind's "giant leap," Lillie was just hoping none of the suit's stitches popped.)

Each spacesuit was custom-made for its wearer. The seamstresses painstakingly stitched together twenty-one layers of gossamer-thin fabric to a tolerance of one sixty-fourth of an inch on the normal sewing machines of the time—there were no special machines for this special task. While these fabric layers gave the wearer some flexibility in movement, and protected him, they also held on to body heat and the heat generated by the machinery in the wearer's backpack, which contained life-support systems including an oxygen pump. This is where the second pump comes in. To combat that heat, engineers designed a separate piece of clothing to be worn inside the main spacesuit. It's known as the Liquid Cooling and Ventilation Garment (or LCVG), but it looks a lot like a version of the kind of onesie worn by babies (and teenagers and comfort-seeking couch potatoes). A form-fitting elastic body suit, the LCVG has over 90 meters of small tubes woven into it. In the backpack, a centrifugal pump (similar to a lawn sprinkler) pushes cooled water from a small tank through these tubes to keep the body temperature within a normal range. As the astronauts move and exert themselves physically, they would be in danger of dehydration or overheating, which could be fatal. The pump brings the warm water into the backpack so it can be cooled and recirculated.

Engineering solutions often find their way into other, unexpected applications—like the wheel did, moving from pottery to transport. The centrifugal pump was first outlined in a fifteenth-century treatise by Renaissance artist and architect Francesco di Giorgio Martini as a potential mud-lifting machine, and was later developed by Denis Papin in 1689 for drainage. I wonder whether either of these visionaries could have ever imagined that their work would, one day, be crucial to human space flight.

Most of us won't experience the rigors of outer space. But a large part of the population will spend time in another situation where pumps play a part in the preservation of life: the maternity hospital.

My own route to childbirth was complicated. Getting pregnant required three rounds of fertility treatment—an invasive and laborious procedure whose eventual success had me alternating between dizzying gratefulness and gasping-for-air anxiety. I couldn't bear the thought of having a miscarriage, which is, sadly, still a very common occurrence, and when I started bleeding a little, about six weeks into my pregnancy, I hurried to the hospital trying not to cry. Thankfully, a scan revealed there was nothing to worry about (except, of course, all the other fears I had) and just over seven months later, bathed in bright lights, and feeling slightly cold because of the air conditioner pumping cool air into the operating room, my daughter entered the world.

I was keen to try breastfeeding my child, but for whatever reason—years of physical and emotional trauma from multiple fertility treatments, a challenging pregnancy, or perhaps because it's just really hard—it turned out to be one of the most difficult things I've ever undertaken. You might say that, for me, breastfeeding sucked.

My nipple felt like a fire was being held to them. Waves of strong pins-and-needles sensations flowed from my breast and all the way down my arm when my ducts were stimulated into letting down the milk. I would sit still, every muscle in my body tensed, tears streaming down my face, while my daughter guzzled. It was agony, but in my vulnerable, hormone-fueled emotional state, with a head full of all the messages about the benefits of breastfeeding, I felt that I had to persist, in order to be a good mother. I spent six to eight hours a day enduring what I can only describe as the worst pain of my life. I couldn't sleep. I couldn't take a shower. There was no break. She cried and I cringed, my body curling in anticipation of the pain. I didn't know it at the time, but I was plummeting into postpartum depression.

I produced plenty of milk, which sounds great in principle, but it meant I was frequently engorged and blocked. My breast became a flat wall that my child couldn't latch on to. The ducts were stretched and sensitive. Very often, they became blocked, and I could feel lumps in my breast, lumps that I could only ease by massaging them—when even just touching them made me cry with pain—or by feeding her. If I didn't, I risked mastitis.

While I was consumed by this fog, my husband reminded me

that a friend had given me a manual breast pump in the months leading up to my daughter's birth. I got it out and fumbled with its parts. A shield shaped like a funnel to envelop my nipple. A bottle that might or might not fill up with milk. A piece with various little valves that connected the two together, to which was attached a lever to be pressed to pull milk out of my breast. In my state of extreme exhaustion and pain, the pump suddenly offered a world of wonderful possibilities. To extract my milk in my own time, at my own pace, with no tiny mouth clamping hard on my nipple. To dream of a stretch of sleep longer than two hours. To enable my husband to feed our daughter. To find some freedom from the relentless and never-ending demand on my body, a body that had already been through so much.

A few weeks in, once my mind felt less foggy, as I sat with my hand rhythmically pressing the lever in front of my breast to create my daughter's next meal, I started wondering how the breast pump had originated and evolved over the decades. (Possibly not what would come to everybody's mind, but what can I say? Breast-pumping enforces a lot of thinking time—and, in the end, I'm a nerdy engineer.) The breast pump has to do some important tasks: it must create enough pressure to draw out milk, but without damaging the delicate tissue of the nipple and breast (so adjustable pressure and speed would be ideal). It needs to be easy to dismantle and easy to clean, so that babies don't fall ill. It needs to mimic the action of a baby's mouth and make sure the supply of milk is maintained. And every breast is so different—how do you design a pump that can adapt?

While the pump gave me much needed flexibility, it did take up a lot of my time, and I felt like a farm animal amid the various contraptions attached to my breast. Turns out, that's no coincidence, as the inspiration for an early breast pump came from milking cows. In the 1890s, inventors John Hartnett and David

Robinson patented a milker for animals in Australia. The machine used a vacuum that pulsated so it would stimulate the udder to release the milk more naturally. This device, however, is no more impersonal and off-putting than most of the other early forms of the breast pump. The briefest of internet explorations will show you eighteenth-century versions beautifully fashioned from brass and glass, but quite clinical and forbidding; an 1897 contraption that most resembles the bulb horn on an old-fashioned bike; and a strange device from the twentieth century with a glass cup that could be attached to a breast, and a pipe that the woman used to suck the milk out (I really can't see how this could possibly have worked).

In 1898, another man, Joseph H. Hoover (who, perhaps thankfully for breastfeeding parents, was not also the inventor of the vacuum cleaner), introduced a "suck and release" function, which used a spring. This was the first time that a mechanism was added that didn't apply a constant pull on the breast, and this step influenced the design of the pumps we still use today. Einar Egnell, a Swedish engineer, designed the first mechanical pump for humans in 1956. He was the first to experimentally calculate the maximum safe pressure on the breast before tissue damage occurs, and, to better mimic a baby, worked out the ideal rate of pulsation (forty-seven times a minute, in case you were wondering).

At this stage, however, breast-pumping was still seen as a last resort—something you might turn to due to the baby's health or because the nipple on the breast was inverted, making it difficult for the baby to latch on to—so designs were cumbersome. Egnell's version was a large, hospital-grade pump intended for babies who were too ill to nurse. The idea of using pumps to ease the pressure on a parent, whether for a break, for convenience, for increasing supply, for building up a store of milk in the freezer—in fact, for anything nonmedical—has only been factored into designs in the last few decades. Personal pumps driven by electricity that you could use in your home only came on the market when I was a teenager, in the late 1990s.

The design of the device itself has hardly changed until recently. Breast pumps generally have a plastic breast shield: the funnel-shaped piece that is placed on the breast with its spout in front of the nipple. Then there is a pump, which can be manual, like mine, or electric. The latter have rotors that change their speed and create suction that varies, pulling the nipple forward, then releasing it to mimic the way a baby drinks. Finally, a bottle collects the milk that emerges from the breast.

I never personally used an electric pump, but both that and the manual version are large and hang off the breast visibly. The pumps in the electrically driven examples are loud. They can be battery-powered to make them portable, but those with the strongest suction need to be plugged into a wall. Air pipes travel to the breast shield and create the pulsing suction. When back at work, parents might end up sitting for long stretches of time alone in rooms that are designated for this purpose (or not) in the office. If you don't pump often enough, you risk becoming engorged and leaking, and also affecting your milk supply. So, anyone who decides to continue breastfeeding once back at work has to constantly manage their diaries and schedule to ensure they can use the pump when needed.

For me, pumping was a game changer, both for my own body's needs and in that it enabled my husband to feed my daughter. But it's important to remember that it's not necessary to be a cis woman or to have been pregnant to breastfeed or to benefit from using a breast pump. Men who are transgender (even after top surgery), adoptive and nonbinary parents, and parents through surrogacy can also often breastfeed, usually via a combination of nipple stimulation and hormonal treatment. Medical advances also mean that, in 2018, a woman who is transgender became the first recorded transgender woman to breastfeed her child, and successfully did so without supplementing with formula for six weeks. The pump can play an important role in all parents' journeys, as a mechanism to stimulate the nipple and therefore the ducts, to produce milk, and I think it's vital that the needs of all types of breastfeeding parents are incorporated into the design.

With the design of breast pumps still stuck in the nineteenth century, and all designed by men, entrepreneurs are now redesigning them by putting the needs of the parent first. One of these is the Elvie pump, which was launched in late 2018. It was conceived by Tania Boler, who is determined to use her company to tackle taboo women's health issues that affect a large proportion of the population but that people are reluctant to talk about openly. I spoke to electronics engineer Shrouk El-Attar (she/they), who worked at Elvie. (I first came across her profile because of her Egyptian belly-dancing drag shows and activism in support of refugees, being one herself.)

The core team developing the Elvie breast pump was composed of around ten people, ranging from researchers to user-experience designers and software and electronics engineers. The idea was to try and erase the picture of what breast pumps had looked like to date and start again from scratch—back to basics. It was the first time that the person using the pump was placed at the center of the design process, and their needs addressed.

Shrouk explained that, based on extensive conversations with new parents who used pumps, the team came up with six core mantras for the design: that it should be silent, hands-free, discreet, smart, easy to use, and, most importantly, not for cows. The result was an egg-shaped pump designed to slip inside the bra. Like the breast pumps before it, it has a shield that sits on the breast. The hub, or pump mechanism, envelops the shield and attaches to it, along with a squat collection bottle. Visually, it increases your cup size by one or two. It is wireless, and very quiet. This means you can attach the pumps to your breasts, put your clothes back on, and then get on with your day, discreetly expressing milk in the office, at meetings, or while on the go. An app on your smartphone shows an estimate of how much milk is collected in the bottle, and you can control the pump speed.

One of the most innovative pieces of engineering in this device is the pump. In other designs, electric pumps used noisy, bulky rotary motors to create changes in air suction. To make a breast pump small enough to fit, at least partially, inside a bra, a different type of mechanism was needed. Shrouk explained that they used a pump called an air pump, which relies on an almost magical phenomenon called piezoelectricity.

Some materials, like quartz crystals, sugar, some ceramics, bone, and even wood, display the piezoelectric effect: pieze being Greek for push. When these materials are squashed or deformed in some way, they generate an internal electrical charge. The opposite is also true; if a voltage is applied to a piezoelectric material, they change shape. Using this principle, engineers have created a small and effective air pump.

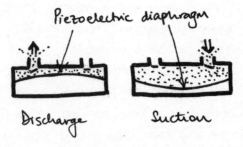

Piezoelectric air pump

A thin circular piece of a piezoelectric material is stuck onto a circular diaphragm made from a flexible material. This arrangement is then put into a case, which has an inlet for air. When a voltage is applied to the system, the piezoelectric layer changes shape and strains, forcing the flexible diaphragm to pull downward and suck air through the inlet. Then, the voltage is adjusted to release the diaphragm. By repeating this motion tens of thousands of times every second, the pump creates an area of low pressure on one side. When the pump stops, the pressure around the device goes back up to normal. In the wearable breast pump, the piezoelectric pump is placed on one side of the breast shield. The shield is made from solid plastic, but with a hole cut out where a flexible silicone diaphragm is connected. Powered by a battery, the suction created by the pump pulls the silicone diaphragm from the breast, in turn creating suction around the nipple. Then the pump pauses to allow release, and the cycle continues, emulating the sucking action of a baby to stimulate the nipple and begin drawing out milk.

There are, of course, pros and cons to the different breast pumps—the hospital-grade pumps that require a connection to a wall are large and unwieldy, but create the strongest suction, which is important for those of us prone to blockages and lumps in our ducts. Large bottles can be attached to them to collect bigger quantities of milk. Wearable pumps are quiet and portable, but don't provide the same levels of suction or capacity for collecting milk. There is also the question of expense and accessibility, and whether you can get a pump on your health insurance or can afford to pay out of your pocket.

Ultimately, based on the shape of your body and your requirements, you can make a choice. That choice exists because, as Shrouk said to me, after 160 years, someone decided that parents who breastfeed shouldn't be treated like cows. Finally, a breast pump has been designed starting with the contemporary needs of families.

It may seem obvious that engineering should consider and consult the users of a product during its conception and development, but as we've seen, this hasn't always been the case. If it had, Paulina Gelman wouldn't have been too short to control the plane, and would have been a pilot. If it had, those consigned to domestic tasks would have had the platform to create and patent dishwashers and other appliances that actually worked at a much earlier date. And if it had, technological advancement would benefit all people, not just those in power.

After all, our engineering becomes a part of our world and shapes our present and future. So, as engineers, let's place the good of the planet and its inhabitants at the heart of our work.

Conclusion

Engineering tells an intrinsically human story.

That's one of the things I found exciting about it when I lived in New York as a little kid and began looking at the skyscrapers surrounding me. At first, of course, I was just bowled over by the sheer scale and drama of these constructions. But gradually, my curiosity deepened. These structures made from brick, stone, and concrete were, I realized, evidence of us: our needs and wants, how we dreamed up and then created solutions, how we went on to organize towns or cities, leading to the shape of how we live our daily lives. From cars to computers to coffee machines, engineering makes our humanity what it is. It's how we interact, with each other and with the planet.

Engineering can reveal who we were. It can tell us our history. Humans have been engineering for millennia, and this ability set us on a different path from other creatures. If we trace this path, much of what we know about our ancient ancestors—from how they ate, built, and lived, to how they formed their societies—is from the flints they shaped, the string they wove, the vehicles in which they traveled, and the instruments they used to investigate and navigate. We can see how, in the past, an object like the wheel, conceived and created in one part of the world, was shared and spread to other parts: a cross-cultural exchange of knowledge prompted by excitement at how a feat of engineering might change life for the better (or, indeed, for the worse: weapons, too, have traveled to and fro along the Silk Road). Our ancestors learned from each other, sometimes taking a technology back to its basics and rebuilding it in a different way, or adapting it to create a different outcome: such as the creation of artificial fibers and steel rope off the back of Neanderthal string.

Engineering also reveals who we are. Just as historical objects can shine a light on our past, so asking questions about the things that surround us now can reveal much about our present. Engineering can appear forbidding, boring, or overwhelming, and even alien, an impenetrable black box. But, given that it's at the heart of how we live and the communities we create, I believe that understanding how the stuff we've made works can be rewarding, even revelatory. (It's no accident that the original "Eureka moment" is attributed to an engineer.) Understanding engineering allows us to learn something about ourselves—examining that black box can be incredibly empowering.

My husband has a cousin called Badrinath. Wherever Badrinath goes, he carries an aura of quiet confidence about him, born of the belief that, even if he doesn't know the solution to a problem, he will be able to figure it out (a quality, alongside the humility to keep learning, that lies at the heart of a great engineer). This belief comes from a lifetime spent exploring how things work, and seeking out simplicity within complexity.

One of Badrinath's earliest memories is of when he was around five years old. His uncle hoisted him up on his shoulders to change a neighbor's light bulb. He lived in a village in southern India in the 1980s with a population of around 20,000, where few homes had electricity, and the unfamiliarity with electricity and the fear of electrocution caused apprehension in the minds of those who did occasionally need a light bulb changed. Young Badrinath was keen to learn, though, and up to the challenge. It turned out to be a light bulb moment in more ways than one, because, from then on, Badrinath was unstoppably curious about technology. When he describes this initial spark that began it all, I think back to my own childhood fascination with destroying crayons—or, I should say, with taking things apart to figure out how they worked. He was, in any case, part of a culture that never threw things away (one that I am also very familiar with, having grown up in metropolitan Mumbai). When appliances or radios stopped working, Badrinath followed his family to one of the repair shops situated on every street corner, and watched as the objects were opened up and the problems diagnosed, and fixed, by self-taught engineers. In high school, he acquired a multimeter, with which he was able to rewire lamps and repair broken blenders. When he moved to the UK, he reconfigured and replumbed his five-bedroom home in Aberdeen to create separate circuits off his boiler, to save energy. Recently, he even dismantled his flatscreen television (much to the concern of his wife and son), figured out it had a faulty board, and replaced it with one he bought online, secondhand.

Apart from the obvious benefits to this outlook—such as saving money and not contributing to the single-use economy—being able to give a product a new life also brings a sense of happiness, satisfaction, and achievement. For Badrinath, his relationship with the object changes; it goes from being something purchased from a shop that he doesn't feel a strong connection to, to something that really belongs to him, that has his mark.

Not being intimidated by the black box of technology can be liberating on an individual level, but beyond this, understanding the nuts and bolts of our objects can lead to an understanding of craft, of design, and also of how we might manage the devastating impact we are having on the planet. And so, engineering can lead us to who we want to be.

Over the course of writing this book, I've had many vigorous, stimulating discussions about the value of breaking things down— and about how breaking apart things (literally and metaphorically) can lead to a better future. One such discussion with a friend, Rebeca Ramos, was particularly illuminating, and sticks in my mind every time I think back over my seven objects and how they work together to build the world in which I live. Rebeca is an architect, artist, and designer on a mission to demystify art and design, making them available to more people, and empowering us to be conscious consumers. Her grandmother worked in a shoe factory in Spain before migrating to Venezuela after the wars, and passed down to her the values of craftsmanship and quality. Rebeca tells me that we only "own" an object for a small proportion of its life, and that having a deeper understanding of design will reveal the massive repercussions, the long chain of events that affect our planet every time we produce or consume something, whether a piece of clothing or an element of a large construction project. Every item requires the creation or extraction of a raw material, then making, packaging, and assembling it—all before we lay our hands on it— and then what happens when we're done with it? The poorer our understanding of what is behind our things, the components that create the whole, the worse our decisions about their quality, value for money, and sustainability will be.

Guru Madhavan, an engineering thought-leader, says that relentlessly producing new objects that may or may not be needed is a questionable act, and that it is paramount to bring ethics, economics, and environmental considerations into the core design of engineering. This ethos is very relevant to the design process that Rebeca believes we should all be attuned to, and that will be a way to stop the relentless production of stuff, which is unsustainable. The UK is one of the largest producers of electronic waste in the world, with nearly 24 kg produced per person in just one year. In the US, this figure was around 21 kg in 2019. It is estimated that around 40 percent of this is exported illegally to other countries to be dumped, where contaminants from these devices can seep into the local food and water supplies. Vast quantities of finite precious metals also end up in landfill.

Thankfully, there are some people stepping up to show us that there is another way. In London, Danielle Purkiss is one such figure. She runs The Big Repair Project, a research project that aims to understand and map the factors that affect the maintenance and repair of electronics across the UK. One of the fundamental concepts of her work is "circular thinking": the idea that every material has a value throughout its lifetime, which should be reconstituted—taken apart and put back together, rather than thrown away. This can apply at the molecular or chemical scale (for example, composting using bacteria or enzymes), or at a more macro physical scale (where you dismantle a device and then repair or replace parts, or break it down so the elements can be reused). Danielle made the important point that there are many products where the majority of harmful emissions are produced during the manufacturing phase of an object, rather than during its use.

Product obsolescence is a big problem, too—that's when products are designed for a limited use-life, or aren't repairable, which leads to them being thrown away. Just 17 percent of consumer electronics (phones, appliances) are recycled in the UK, and this drops to 15 percent in the US. Part of the problem is that such pieces of tech usually contain a complex configuration of materials that are difficult to extract in the first place—and then, if they are not recycled in a controlled setting, they can be damaging to health and environment. The greater the complexity in the design and materials of an object, the harder it is to separate them out again, meaning they often end up in landfill. This is why it's vital to consider how to disassemble a product right from the start of the design process, so that products can first be repaired or upgraded to last longer, and eventually be recycled.

Bringing together the principles of civic engineering, conscious consumption, and knowledge of how our stuff works can only be a good thing, but we also need manufacturers themselves to get on board, as well as policy and government. Overall, Danielle is optimistic. During the last few years, she's seen a rise in community repair hubs, where people help each other fix things, and an increase in advertising for refurbished phones and secondhand tech. Lockdowns in response to the global pandemic forced remote working and study for people who may not have had the tools they needed, so there has been a leap in creating affordable electronics.

By breaking things down and delving into the small, hopefully I've illustrated that far from being overwhelming and cold, engineering—past, present, and future—is stimulating, empowering, and human. In its complexity, you can chip away to find simplicity; but sometimes, that simplicity is deceptive, and its story and science weave together, to take us on a fascinating journey back into the complex. In addition to the benefits we gain as individuals when it is demystified, engineering shows us a path forward that we can take as a species: one with more empathy for all living beings and the planet we inhabit. To take some early steps, to learn about this path and the technology that comes with it, we can start small. We can open up the ostensibly inscrutable and seek to understand the elements from which it is made. And then, before reassembling those elements, we can ask ourselves the question—how might we do it better?

ACKNOWLEDGMENTS

This book was written over two years during the global pandemic, a challenging time for everyone. So first, my thanks go to all the scientists, engineers, healthcare professionals—and everyone else who toiled in the background, often unacknowledged—to keep the world ticking through to the (almost) other side.

Then, my thanks go to:

Patrick Walsh, the best agent, for having faith in me that I would come up with an idea and write it, possibly even well, for all the comments and support, and for being there always.

Kirty Topiwala, my visionary editor at Hodder & Stoughton, who helped me hone the pitch until we nailed it, and for believing that we had a solid book in us. Anna Baty for taking up the baton, and then Izzy Everington, as the babies continued to appear! Izzy, your feedback was invaluable, and you brought it home to the finish line amid all my Gantt charts and spreadsheets, with some pretty hard work.

Quynh Do and John Glusman for believing in the book, your thoughtful feedback, and for bringing my words stateside.

Pascal Cariss, for once again making my sentences sizzle. It's always a pleasure to work with you, and learn so much from you. You are the Pun Master.

Tara O'Sullivan and Nancy Palmquist for your incredibly meticulous work correcting my typos and grammar.

To every single one of the over eighty of you (listed in the Contributor section) who gave me your time and knowledge through video calls, emails, and the occasional in-person meeting, during a really difficult time. You didn't have to speak to me; it was your generosity that allowed me to put this book together. You were my lifeline in our collective isolation; I am deeply, deeply grateful.

My dear friends at Neuwrite, in particular, Subhadra Das, Hana Ayoob, Rebecca Struthers, and Alex O'Brien, for reading chapters and extracts and giving me the nudges I needed to keep going. I can't wait to read all your books. Fatin Marini, Lota Erinne, and Antara Dutt for reading the sections that were well beyond my comfort zone, and helping me enrich the text.

For my worldwide family who existed as faces on screens during this process, your encouragement and presence means everything to me.

My parents, Hem and Lynette, my sister Pooja, my brother-in-law Daniel, and my little Kiah. For all your love.

For my triangle family—little Zarya and Badri, I have no words. We've been through it together, all the ups and downs, and I'm very lucky to have you.

Roma Agrawal
March 2023

CONTRIBUTORS

Thank you to the following for so generously sharing your knowledge and wisdom during the course of writing this book.

Nail

Agnes Jones, *artist and blacksmith*

Andrew Smith, *Rolls-Royce plc*

Azby Brown, *Japanese carpentry expert*

Dr. Bill Eccles, *Bolt Science*

Branca Pegado, *senior architect at Article 25*

Dr. Coralie Acheson, *Arup*

Dr. Dan Ridley-Ellis, *Edinburgh Napier University*

Darren James, *Blue Bear Systems*

Diana Davis, *ACR, National Museum of the Royal Navy*

Dr. Eleanor Schofield, *Mary Rose Trust*

Gervais Sawyer, *FIMMM, wood consultant*

Ian Firth, *FREng, Ian Firth Structural Engineering*

Professor Jan Bill, PhD, *Museum of Cultural Heritage, University of Oslo*

Jean-Michel Munn, *The Shuttleworth Collection*

Dr. John Roberts, *FREng, Jacobs*

Dr. Julian Whitewright, *maritime archaeologist*

Mitch Peacock, BEng (Hons), *woodworking tutor and author*

Morgan Creed, *MSc BA (Hons), National Museum of the Royal Navy*

Nicola Grahamslaw, *SS Great Britain Trust*

Omar Sharif, BEng (Hons), *CEng, MIStructE*

Rebecca Wilton (She/Her), *MA PGCert PGDip BA (Hons), The Ladybirdhouse*

Rich Maynard, *Much Hadham Forge*

Steve Hyett, *Eliza Tinsley*

Wheel

Darren Ellis, *Darren Ellis Pottery / Institute of Making*

Greg Rowland, *Mike Rowland & Son*

Mark Sanders, *MAS Design Products Ltd.*

Robert Hurford

Will Stanczykiewicz, *NASA Johnson Space Center*

Spring

Adam Fox, *CEng AMIOA, Mason UK*

Buma, *Mongolian Archery*

Doug Valerio, *Mason Industries*

Goyo Reston, *Goyo Travel*

James Beer, *Arup*

Jens Nielsen, *f2c*

Jordi Femenia, *Mason UK*

Keith Scobie-Youngs, *Cumbria Clock Company*

Kiran Shekar, *Minutia Repeater*

Martin Raisborough, *BEng (Hons), MIOA*

Michael Wolf, *GERB Schwingungsisolierungen GmbH & Co. KG*

Nicoletta Galluzzi, *MSc CEng MICE, structural engineer*

Dr. Nikhil Mistry

Oliver Farrell, *CEng Meng FIMechE SIA, Farrat*

Dr. Rebecca Struthers, *Struthers Watchmakers*

Roger Kelly, *building isolation specialist*

Stefan Haberl, *translator (technical, patents)*

Magnet

Dr. Andrew Princep (He/Him), *Marketcast*

Dr. Eleanor Armstrong, *Stockholm University*

Gavin Payne, *The Old Telephone Company*

Professor Hidenori Mimura, *Shizuoka University*

Keith Rhodes, *Magnetic Products Inc.*

Dr. Suvobrata Sarkar, *Rabindra Bharati University*

Dr. Suzie Sheehy, *University of Melbourne*

Lens

Ben Pipe, *Ben Pipe Photography*

Brian J. Ford, *author and broadcaster*

Dr. Ceri Brenner, *ANSTO Centre for Accelerator Science*

Christiana Antoniadou Stylianou, *BSc (Hons) MSc MPhil, senior clinical embryologist*

David Noton, *David Noton Photography*

Dr. Geoff Belknap, *Science and Media Museum*

Hanan Dowidar, *1001 Inventions*

Kenneth Sanders, *BSc DSc (Hons) CCMI, Worshipful Company of Scientific Instrument Makers*

Dr. Kwasi Kwakwa, *Sanger-EBI*

Dr. Michael Pritchard, *Royal Photographic Society, Bristol, UK*

Professor Mohamed El Gomati OBE, *BSc DPhil FIoP FRMS, University of York*

Phillip Roberts

Professor Stanley Botchway, *UKRI–Science and Technology Facilities Council*

String

Harjit Shah, *JAS Musicals*

Helen Sheldon (She/Her), *BSc CEng MIOA MWES FRSA, RBA Acoustics*

Karen Yates, *Macalloy*

Mark Ellis, *Macalloy*

Mary Lewis, *Heritage Crafts*

Professor Rachel Worth (She/Her), *BA (Hons) (Cantab) PGCE MA PhD, Arts University Bournemouth*

Toss Levy, *Indian Musical Instruments*

Pump

Dr. Clara Barker (She/Her), *MRSC, MInstP, Oxford University & Linacre College*

Dallas Campbell, *broadcaster and author*

Dr. Rohin Francis, *Colchester Hospital and the Essex Cardiothoracic Centre*

Shrouk El-Attar (She/They), *Shrouk El-Attar Consulting*

Vinita Marwaha Madill (She/Her), *Rocket Women*

General

Dr. Ainissa Ramirez, *scientist and author*

Badrinath Hebsur

Danielle Purkiss, *University College London*

Rebeca Ramos, *Studio RARE*

Thanks also to Jemima Waters, *Folkies Music, Chiaro Technology, the Thomas Jefferson Foundation, British and Irish Association of Fastener Distributors, The Golden Hinde, the Institution of Structural Engineers, the Institution of Civil Engineers, The Royal Society, and Wellcome Collection.*

BIBLIOGRAPHY AND FURTHER READING

General

Earth911. "20 Staggering E-Waste Facts in 2021." Earth911, October 11, 2021.

Forman, Chris, and Claire Asher. *Brave Green World*. MIT Press, 2021.

Gadd, Karen. *TRIZ for Engineers: Enabling Inventive Problem Solving*. Wiley, 2011.

Holmes, Keith C. *Black Inventors: Crafting Over 200 Years of Success*. Global Black Inventor Research Projects Inc., 2008.

Jaffe, Deborah. *Ingenious Women: From Tincture of Saffron to Flying Machines*. Sutton Publishing, 2004.

Jayaraj, Nandita, and Aashima Freidog. *31 Fantastic Adventures in Science: Women Scientists of India*. Puffin Books, 2019.

Madhavan, Guru. *Think Like an Engineer: Inside the Minds That Are Changing Our Lives*. Oneworld, 2016.

Malloy, Kai. "UK on Track to Become Europe's Biggest e-Waste Contributor." *Resource*, October 21, 2021.

McLellan, Todd. *Things Come Apart 2.0*. Thames & Hudson, 2013.

Petroski, Henry. *The Evolution of Useful Things: How Everyday Artifacts—from Forks and Pins to Paper Clips and Zippers—Came to Be as They Are*. Vintage Books, 1992.

Rattray Taylor, Gordon, ed. *The Inventions That Changed the World: An Illustrated Guide to Man's Practical Genius Through the Ages*. Reader's Digest, 1983.

Toner Buzz. "Staggering E-Waste Facts & Statistics 2022," March 9, 2022.

Walker, Robin. *Blacks and Science Volume 2: West and East African Contributions to Science and Technology.* Reklaw Education, 2016.

Walker, Robin. *Blacks and Science Volume 3: African American Contributions to Science and Technology.* Reklaw Education, 2013.

Nail

Ackroyd, J. A. D. "The Aerodynamics of the Spitfire." *Journal of Aeronautical History,* 2016.

Alexievich, Svetlana. *The Unwomanly Face of War.* Penguin Random House, 1985.

Anne of All Trades. "Blacksmithing: Forging a Nail by Hand." YouTube, May 17, 2019.

Atack, D., and D. Tabor. "The Friction of Wood." *Proceedings of the Royal Society A,* vol. 246, no. 1247, August 26, 1958.

Bill, Jan. "Iron Nails in Iron Age and Medieval Shipbuilding." In *Crossroads in Ancient Shipbuilding.* Roskilde, 1991.

Budnik, Ruslan. "Instrument of the Famous 'Night Witches,'" War History Online, August 8, 2018.

Castles, Forts and Battles. "Inchtuthil Roman Fortress."

Chervenka, Mark. "Nails as Clues to Age." Real or Repro.

Collette, Q. "Riveted Connections in Historical Metal Structures (1840–1940). Hot-Driven Rivets: Technology, Design and Experiments." 2014.

Collette, Q., I. Wouters, and L. Lauriks. "Evolution of Historical Riveted Connections: Joining Typologies, Installation Techniques and Calculation Methods." *Structural Studies, Repairs and Maintenance of Heritage Architecture XII,* 2011, pp. 295–306.

Collins, W. H. "A History 1780–1980." Swindell and Co., From Eliza Tinsley & Co. Ltd.

Corlett, Ewan. *The Iron Ship—The Story of Brunel's SS Great Britain.* Conway Maritime Press, 2002.

Dalley, S. *The Mystery of the Hanging Garden of Babylon: An Elusive World Wonder Traced.* OUP Oxford, 2013.

Eliza Tinsley. "The History of Eliza Tinsley."

Eliza Tinsley & Co. Ltd. "Nail Mistress." [Eliza Tinsley Obituary].

The Engineering Toolbox. "Nails and Spikes—Withdrawal Force."

Essential Craftsman. "Screws: What You Need to Know." YouTube, June 27, 2017.

Fastenerdata. "History of Fastenings."

Formisano, Bob. "How to Pick the Right Nail for Your Next Project." *The Spruce*, January 11, 2021.

Forest Products Laboratory. *Wood Handbook: Wood as an Engineering Material*. United States Department of Agriculture, 2010.

Founders Online. "From Thomas Jefferson to Jean Nicolas Démeunier, 29 April 1795." University of Virginia Press.

Glasgow Steel Nail. "The History of Nail Making."

Goebel Fasteners. "History of Rivets & 20 Facts You Might Not Know." October 15, 2019.

Hening, W. W. *The Statutes at Large: Being a Collection of All the Laws of Virginia, from the First Session of the Legislature, in the Year 1619 : Published Pursuant to an Act of the General Assembly of Virginia, Passed on the Fifth Day of February One Thousand Eight Hundred and Eight*. 1823.

How, Chris. "The British Cut Clasp Nail." In *Proceedings of the First Construction History Society Conference, Queens' College, University of Cambridge*, Construction History Society, 2014.

How, Chris. *Early Steps in Nail Industrialisation*. Queens' College, University of Cambridge, 2015.

How, Chris. "Evolutionary Traces in European Nail-Making Tools." In *Building Knowledge, Constructing Histories*, CRC Press, 2018.

How, Chris. *Historic French Nails, Screws and Fixings: Tools and Techniques*. Furniture History Society of Australasia, 2017.

How, Chris. "The Medieval Bi-Petal Head Nail." In *Further Studies in the History of Construction: The Proceedings of the Third Annual Conferences of the Construction History Society*, Construction History Society, 2016.

Hunt, Kristen. "Design Analysis of Roller Coasters." Thesis submitted to Worcester Polytechnic Institute, May 2018.

Inspectapedia. "Antique Nails: History & Photo Examples of Old Nails Help Determine Age & Use."

Johnny from Texas. "Builders of Bridges (1928) Handling Hot Rivets." YouTube, February 23, 2020.

Jon Stollenmeyer, Seek Sustainable Japan. "Love of Japanese Architecture + Building Traditions." YouTube, October 15, 2020.

Kershaw, Ian. "Before Nails, There Was Pegged Wood Construction." *Outdoor Revival*, April 14, 2019.

Mapelli, C., R. Nicodemi, R. F. Riva, M. Vedani, and E. Gariboldi. "Nails of the Roman Legionary." *La Metallurgia Italiana*, 2009.

Morgan, E. B., and E. Shacklady. *Spitfire: The History*. Key Publishing, Stamford, 1987.

Much Hadham Forge Museum. "Our Museum."

Museum of Fine Arts Boston. "Jug with Lotus Handle."

Neuman, Scott. "Aluminum's Strange Journey from Precious Metal to Beer Can." NPR, December 10, 2019.

Nord Lock. "The History of the Bolt."

Perkins, Benjamin. "Objects: Nail Cutting Machine, 1801, by Benjamin Perkins. M29 [Electronic Edition]." Massachusetts Historical Society.

Pete & Sharon's SPACO. "Making Hand Forged Nails."

Pitts, Lynn F., and J.K. St Joseph. *Inchtuthil: The Roman Legionary Fortress Excavations, 1952–65*. Society for the Promotion of Roman Studies, 1985.

Rivets de France. "History."

Roberts, J. M. "The 'PepsiMax Big One' Rollercoaster Blackpool Pleasure Beach." *The Structural Engineer*, vol. 72, no. 1, 1994.

Roberts, John. "Gold Medal Address: A Life of Leisure." *The Structural Engineer*, June 20, 2006.

Rybczynski, Witold. *One Good Turn: A Natural History of the Screwdriver & the Screw*. Scribner, 2001.

Sakaida, Henry. *Heroines of the Soviet Union 1941–1945*. Osprey Publishing, 2003.

Sedgley Manor. "Black Country Nail Making Trade."

Shuttleworth Collection. "Solid Riveting Procedures." [Design guidance].

Sullivan, Walter. "The Mystery of Damascus Steel Appears Solved." *The New York Times*, September 29, 1981.

Tanner, Pat. "Newport Medieval Ship Project: Digital Reconstruction and Analysis of the Newport Ship." [3D Scanning Ireland]. May 2013.

Taylor, Jonathan. "Nails and Wood Screws." *Building Conservation*.

Thomas Jefferson's Monticello. "Nailery."

TR Fastenings. "Blind Rivet Nuts, Capacity Tables." [Company Brochure, Edition 2].

Truini, Joseph. "Nails vs. Screws: How to Know Which Is Best for Your Project." *Popular Mechanics*, March 29, 2022.

Twickenham Museum. "Henrietta Vansittart, Inventor, Engineer and Twickenham Property Owner."

Visser, Thomas D. *A Field Guide to New England Barns and Farm Buildings.* University Press of New England, 1997.

Wagner Tooling Systems. "The History of the Screw." [Company brochure].

Weincek, Henry. "The Dark Side of Thomas Jefferson." *Smithsonian Magazine*, October 10, 2012.

Willets, Arthur. *The Black Country Nail Trade.* Dudley Leisure Services, 1987.

Wilton, Rebecca. "The Life and Legacy of Eliza Tinsley (1813–1882), Black Country Nail Mistress." MA in West Midlands History, University of Birmingham.

Winchester, Simon. *The Perfectionists: How Precision Engineers Created the Modern World.* HarperCollins, 2018.

Zhan, M., and H. Yang. "Casting, Semi-Solid Forming and Hot Metal Forming." In *Comprehensive Materials Processing,* Elsevier, 2014.

Wheel

American Physical Society. "On the Late Invention of the Gyroscope." *Bulletin of the American Physical Society*, vol. 57, no. 3.

Anthony, David W. *The Horse, the Wheel, and Language: How Bronze-Age Riders from the Eurasian Steppes Shaped the Modern World.* Erenow.

Art-A-Tsolum. "4,000 Years Old Wagons Found in Lchashen, Armenia." December 28, 2017.

Baldi, J. S. "How the Uruk Potters Used the Wheel." EXARC, YouTube, 2020.

BBC News. "Stone Age Door Unearthed by Archaeologists in Zurich." October 21, 2010.

Belancic Glogovcan, Tanja. "World's Oldest Wheel Found in Slovenia." I Feel Slovenia, January 6, 2020.

Bellis, Mary. "The Invention of the Wheel." ThoughtCo, December 20, 2020.

Berger, Michele W. "How the Appliance Boom Moved More Women into the Workforce." *Penn Today*, January 30, 2019.

Bowers, Brian. "Social Benefits of Electricity." *IEE Proceedings A (Physical Science, Measurement and Instrumentation, Management and Education, Reviews)*, vol. 135, no. 5, May 5, 1988.

Brown, Azby. *The Genius of Japanese Carpentry: Secrets of an Ancient Craft.* Tuttle, 2013.

Burgoyne, C. J., and R. Dilmaghanian. "Bicycle Wheel as Prestressed Structure." *Journal of Engineering Mechanics*, vol. 119, no. 3, March 1993.

Cassidy, Cody. "Who Invented the Wheel? And How Did They Do It?' *Wired*.

Chariot VR. "A Brief History of the Spoked Wheel."

Cochran, Josephine G. "Dish Washing Machine." United States Patent Office 355, 139, issued December 28, 1886.

Davidson, L.C. *Handbook for Lady Cyclists*. Hay Nisbet, 1896.

Davis, Beverley. "Timeline of the Development of the Horse." *Sino-Platonic Papers*, no. 177, August 2007.

Deloche, Jean. "Carriages in Indian Iconography." In *Contribution to the History of the Wheeled Vehicle in India*, 13–48. Français de Pondichéry, 2020.

Deneen Pottery. "Pottery: The Ultimate Guide, History, Getting Started, Inspiration."

Deutsches Patent- und Markenamt. "Patent for Drais' 'Laufmaschine,' The ancestor of all bicycle."

engineerguy. "How a Smartphone Knows Up from Down (Accelerometer)." YouTube, May 22, 2012.

European Space Agency. "Gyroscopes in Space."

Evans-Pughe, Christine. "Bold Before Their Time." *Engineering and Technology Magazine*, June 2011.

Freeman's Journal and Daily Commercial Advertiser, August 30, 1899.

Gambino, Megan. "A Salute to the Wheel." *Smithsonian Magazine*, June 17, 2009.

Garcia, Mark. "Integrated Truss Structure." September 20, 2018.

Garis-Cochran, Josephine G. "Advertisement for Dish Washing Machine." 1895.

Gibbons, Ann. "Thousands of Horsemen May Have Swept into Bronze Age Europe, Transforming the Local Population." *Science*, February 21, 2017.

Glaskin, Max. "The Science Behind Spokes." *Cyclist*, April 28, 2015.

Green, Susan E. *Axle and Wheel Terminology, an Historical Dictionary*.

Haan, David de. *Antique Household Gadgets and Appliances c. 1860 to 1930.* Blandford Press, 1977.

Harappa. "Chariots in the Chalcolithic Rock Art of India."

Hazael, Victoria. "200 Years Since the Father of the Bicycle Baron Karl von Drais Invented the 'Running Machine.'" *Cycling UK*.

History Time. "The Nordic Bronze Age / Ancient History Documentary." YouTube, February 22, 2019.

ISS Live! "Control Moment Gyroscopes: What Keeps the ISS from Tumbling through Space?" NASA.

Kenoyer, J. M. "Wheeled Vehicles of the Indus Valley Civilization of Pakistan and India," University of Wisconsin–Madison. January 7, 2004.

Kessler, P. L. "Kingdoms of the Barbarians—Uralics." History Files.

Lemelson. "Josephine Cochrane: Dish Washing Machine."

Lewis, M. J. T. "Gearing in the Ancient World." *Endeavour*, vol. 17, no. 3, January 1, 1993.

Lloyd, Peter. "Who Invented the Toothed Gear?" Idea Connection.

Manners, William. *Revolution: How the Bicycle Reinvented Modern Britain*. Duckworth, 2019.

Manners, William. "The Secret History of 19th Century Cyclists." *Guardian*, June 9, 2015.

Minetti, Alberto E., John Pinkerton, and Paola Zamparo. "From Bipedalism to Bicyclism: Evolution in Energetics and Biomechanics of Historic Bicycles." *Proceedings of the Royal Society of London. Series B: Biological Sciences*, vol. 268, no. 1474, July 7, 2001.

NASA. "Reference Guide to the International Space Station." September 2015.

NASA. "International Space Station Familiarization: Mission Operations Directorate Space Flight Training Division." July 31, 1998.

NASA History Division. "EP—107 Skylab: A Guidebook."

NASA Video. "Gyroscopes." YouTube, May 22, 2013.

Pollard, Justin. "The Eccentric Engineer." *Engineering and Technology Magazine*, July 2018.

Postrel, Virginia. "How Job-Killing Technologies Liberated Women." Technology & Ideas: Bloomberg, March 14, 2021.

Quora. "How Does the International Space Station Keep Its Orientation?" *Forbes*, April 26, 2017.

Racing Nellie Bly. "Chipped China Inspired Josephine Cochrane to Invent Effective Victorian Era Dishwashers." November 12, 2017.

Schaeffer, Jacob Christian. *Die bequeme und höchstvortheilhafte Waschmaschine*, 1767.

ScienceDaily. "Fridges and Washing Machines Liberated Women, Study Suggests."

ScienceDaily. "Reinventing the Wheel—Naturally."

Simply Space. "ISS Attitude Control—Torque Equilibrium Attitude and Control Moment Gyroscopes." YouTube, September 6, 2019.

Sommeria, Joël. "Foucault and the Rotation of the Earth." *Science in the Making: The Comptes Rendus de l'Académie des Sciences Throughout History*, vol. 18, no. 9, November 1, 2017.

Stockhammer, Philipp W., and Joseph Maran, eds. *Appropriating Innovations: Entangled Knowledge in Eurasia, 5000–1500 BCE*. Oxbow Books, 2017.

Sturt, George. *The Wheelwright's Shop*. Cambridge University Press, 1923.

Tietronix. "Console Handbook: ADCO Attitude Determination and Control Officer." [Technical Handbook prepared for NASA, Johnson's Space Centre].

Tucker, K., N. Berezina, S. Reinhold, A. Kalmykov, A. Belinskiy, and J. Gresky. "An Accident at Work? Traumatic Lesions in the Skeleton of a 4th Millennium BCE 'Wagon Driver' from Sharakhalsun, Russia." *HOMO*, vol. 68, no. 4, August 2017.

United States Patent and Trademark Office. "Josephine Cochran: 'I'll Do It Myself.'"

Vogel, Steven. *Why the Wheel Is Round: Muscles, Technology, and How We Make Things That Move*. University of Chicago Press, 2018.

Wolchover, Natalie. "Why It Took So Long to Invent the Wheel." *Live Science*, March 2, 2012.

Woodford, Chris. "How Do Wheels Work? Science of Wheels and Axles." Explain That Stuff, January 27, 2009.

Wright, John, and Robert Hurford. "Making a Wheel—How to Make a Traditional Light English Pattern Wheel." Rural Development Commission, 1997.

Spring

American Physical Society. "June 16, 1657: Christiaan Huygens Patents the First Pendulum Clock." June 2017.

American Physical Society. "March 20, 1800: Volta Describes the Electric Battery." March 2006.

Andrewes, William J. H. "A Chronicle of Timekeeping." *Scientific American*, February 1, 2006.

Animagraffs. "How a Mechanical Watch Works." YouTube, November 20, 2019.

"Antiquarian Horology." *The Athenian Mercury VI*, no. 4, query 7, February 13, 1692/93.

Arbabi, Ryan. "At the Extremes of Acoustic Science." [Conference Paper]. Farrat, July 2021.

ArchDaily. "House of Music/Coop Himmelb(l)Au." April 14, 2014.

Archery Historian. "Mongolian Bow VS English Longbow—Advantages and Drawbacks." June 23, 2018.

Automated Industrial Motion. "ALL ABOUT SPRINGS: Comprehensive Guide to the History, Use and Manufacture of Coiled Springs." 2019.

Backwell, Lucinda, Justin Bradfield, Kristian J. Carlson, Tea Jashashvili, Lyn Wadley, and Francesco d'Errico. "The Antiquity of Bow-and-Arrow Technology: Evidence from Middle Stone Age Layers at Sibudu Cave." *Antiquity*, vol. 92, no. 362, April 2018.

BBC News. "A Point of View: How the World's First Smartwatch Was Built." September 27, 2014.

Beacock, Ian P. "A Brief History of (Modern) Time." *The Atlantic*, December 22, 2015.

Beever, Jason Wayne, and Zoran Pavlovic. "The Modern Reproduction of a Mongol Era Bow Based on Historical Facts and Ancient Technology Research." EXARC, June 1, 2017.

Bellis, Mary. "The History of Mechanical Pendulum and Quartz Clocks." ThoughtCo, April 12, 2018.

Berman, Mark, et al. "The Staggering Scope of U.S. Gun Deaths Goes Far Beyond Mass Shootings." *Washington Post*, July 8, 2022.

Blakemore, Erin. "Who Were the Mongols?" *National Geographic*, June 21, 2019.

Blumenthal, Aaron, and Michael Nosonovsky. "Friction and Dynamics of Verge and Foliot: How the Invention of the Pendulum Made Clocks Much More Accurate." *Applied Mechanics*, vol. 1, no. 2, April 29, 2020.

Britannica. "Bow and Arrow."

Brown, Emily Lindsay. "The Longitude Problem: How We Figured Out Where We Are." *The Conversation*, July 18, 2013.

Brown, Erik. "How the Ancients Improved Their Lives with Archery." *Medium*, October 15, 2020.

Brownstein, Eric X. "The Path of the Arrow: The Evolution of Mongolian National Archery." World Learning/SIT Study Abroad, Mongolia, Spring 2008.

Buckley Ebrey, Patricia. "Crossbows." *A Visual Sourcebook of Chinese Civilization*.

Bues, Jon. "Introducing: The Zenith Defy 21 Ultraviolet." *Hodinkee*, June 1, 2020.

Burgess, Ebenezer. *Surya Siddhanta Translation*. Internet Archive.

Cartwright, Mark. "Crossbows in Ancient Chinese Warfare." *World History Encyclopedia*, July 17, 2017.

Charles Frodsham and Co Ltd. "Discovering Harrison's H4."

Chong, Alvin. "In-Depth: Time Consciousness and Discipline in the Industrial Revolution." SJX Watches, July 21, 2020.

Croix Rousse Watchmaker. "Explanation, How Verge Escapement Works." YouTube, September 11, 2017.

Currie, Neil George Roy. *Kinky Structures.* School of Computing, Science and Engineering, University of Salford, 2020.

Daltro, Ana Luiza. "Interview. Yasuhisa Toyota, the Sound Wizard." *ArchiExpo e-Magazine*, February 12, 2018.

Davies, B. J. "The Longevity of Natural Rubber in Engineering Applications." The Malaysian Rubber Producers Research Association, reprinted from article in *Rubber Developments*, vol. 41, no. 4.

DeVries, Kelly, and Robert Douglas Smith. *Medieval Weapons: An Illustrated History of Their Impact.* ABC-CLIO, 2007.

Einsmann, Scott. "History Proves Archery's Roots Are Ancient, and This Evidence Is Awesome!" *Archery 360*, May 3, 2017.

Fact Monster. "Accurate Mechanical Clocks." February 21, 2017.

Farrat. "Acoustic Isolation of Concert Halls." [Design Guidance].

Farrat. "Building Vibration Isolation Systems: Vibration Control for Buildings and Structures." [Design Guidance].

Farrell, Oliver. "From Acoustic Specification to Handover. A Practical Approach to an Effective and Robust System for the Design and Construction of Base (Vibration) Isolated Buildings." *Farrat*, 2017.

Farrell, Oliver, and Ryan Arbabi. "Long-Term Performance of Farrat LNR Bearings for Structural Vibration Control."

Fowler, Susanne. "From Working on Watches to Writing About Them." *New York Times*, September 8, 2021.

Fusion. "A Briefer History of Time: How Technology Changes Us in Unexpected Ways." YouTube, February 18, 2015.

GERB. "Floating Floors and Rooms." [Company Sales Brochure], 2016.

GERB. "Tuned Mass Dampers for Bridges, Floors and Tall Structures." [Technical Paper].

GERB. "Vibration Control Systems Application Areas." [Sales Brochure].

GERB. "Vibration Isolation of Buildings." [Sales Brochure].

Gledhill, Sean. "Pushing the Boundaries of Seismic Engineering." *The Structural Engineer*, vol. 89, no, 12, June 21, 2011.

Glennie, Paul, and Nigel Thrift. "Reworking E. P. Thompson's 'Time, Work—Discipline and Industrial Capitalism .'" *Time & Society*, vol. 5, no. 3, October 1996.

Gonsher, Aaron. "Interview: Master Acoustician Yasuhisa Toyota." *Red Bull Music Academy Daily*, April 14, 2017.

Graceffo, Antonio. "Mongolian Archery: From the Stone Age to Naadam." *Bow International*, August 14, 2020.

HackneyedScribe. "Han Dynasty Crossbow III." History Forum, July 2, 2019.

Harder, Jeff, and Sharise Cunningam. "Who Invented the First Gun?" HowStuff-Works, January 12, 2011.

Harris, Colin S., ed. *Engineering Geology of the Channel Tunnel*. American Society of Civil Engineers, 1996.

Hirst, Kris. "The Invention of Bow and Arrow Hunting Is at Least 65,000 Years Old." ThoughtCo, May 19, 2019.

Hunt, Hugh. "Inside Big Ben: Why the World's Most Famous Clock Will Soon Lose Its Bong." *The Conversation*, April 29, 2016.

Institute for Health Metrics and Evaluation. "Six Countries in the Americas Account for Half of All Firearm Deaths." August 24, 2018.

Kaveh, Farrokh, and Manouchehr Moshtagh Khorasani. "The Mongol Invasion of the Khwarazmian Empire: The Fierce Resistance of Jalal-e Din." *Medieval Warfare*, vol. 2, no. 3, 2012.

Landes, David S. *Revolution in Time: Clocks and the Making of the Modern World*. Harvard University Press, 1998.

Loades, Mike. *The Crossbow*. Osprey Publishing, 2018.

Lombardi, Michael. "First in a Series on the Evolution of Time Measurement: Celestial, Flow, and Mechanical Clocks [Recalibration]." *IEEE Instrumentation & Measurement Magazine*, vol. 14, no. 4, August 2011.

Mason Industries. "ASHRAE Lecture: Noise and Vibration Problems and Solutions." November 1966.

Mason Industries. "BUILDING ISOLATION SLFJ Spring Isolators BBNR Rubber Isolation Bearings."

Mason Industries. "History."

Mason Industries. "Double Deflection Neoprene Mount." BULLETIN ND-26-1. [Technical Paper].

Mason Industries. "Mason Jack-Up Floor Slab System." 2017.

Mason Industries. "Spring Mount for T.V. Studio Floor for Columbia Broadcasting System." [Product Specification].

Mason Industries, Inc. "FREE STANDING SPRING MOUNTS and HEIGHT SAVING BRACKETS." [Product Specification], 2017.

Mason UK. "Concrete Floating Floor Vibration Isolation, House of Music, Denmark."

Mason UK Ltd—Floating Floors, Vibration Control & Acoustic Products. "Seismic Table Testing of Inertia Base Frame | DCL Labs." YouTube, April 23, 2020.

May, Timothy. *The Mongol Art of War*. Casemate Publishers, 2007.

McFadden, Christopher. "Mechanical Engineering in the Middle Ages: The Catapult, Mechanical Clocks and Many More We Never Knew About." *Interesting Engineering*, April 28, 2018.

Mills, Charles W. "The Chronopolitics of Racial Time." *Time & Society*, vol. 29, no. 2, May 2020.

Myers, Joe. "In 2016, half of all gun deaths occurred in the Americas." World Economic Forum, August 6, 2019.

The Naked Watchmaker. "Rebecca Struthers."

The National Museum of Mongolian History. "The Mongol Empire of Chingis Khan and His Successors."

North, James David. *God's Clockmaker: Richard of Wallingford and the Invention of Time*, 2005.

Ogle, Vanessa. *The Global Transformation of Time*. Harvard University Press, 2015.

Open Culture. "How Clocks Changed Humanity Forever, Making Us Masters and Slaves of Time." February 19, 2015.

Pearce, Adam, and Jac Cross. "Structural Vibration—a Discussion of Modern Methods." *The Structural Engineer*, vol. 89, no. 12, June 21, 2011.

Physics World. "A Brief History of Timekeeping," November 9, 2018.

Ramboll Group. "House of Music: Harmonic Interaction in Architectural Playground."

Roberts, Alice. "A True Sea Shanty: The Story behind the Longitude Prize." *Observer*, May 17, 2014.

Royal Museums Greenwich. "Longitude Found—the Story of Harrison's Clocks."

Royal Museums Greenwich. "Time to Solve Longitude: The Timekeeper Method." September 29, 2014.

Ruderman, James. "High-Rise Steel Office Buildings in the United States." *The Structural Engineer*, vol. 43, no. 1, 1965.

Saito, Daisuke, Mott MacDonald, and Kazumi Terada. "More for Less in Seismic Design for Bridges—an Overview of the Japanese Approach." *The Structural Engineer*, February 1, 2016.

Salisbury Cathedral. "What Is the Story Behind the World's Oldest Clock?" YouTube, June 4, 2020.

Sample, Ian. "Eureka! Lost Manuscript Found in Cupboard." *Guardian*, February 9, 2006.

Stanley, John. "How Old Is the Bow and Arrow?" *World Archery*, April 8, 2019.

Stilken, Alexander. "Masters of Sound." Porsche Newsroom, November 15, 2017.

Szabo, Christopher. "Ancient Chinese Super-Crossbow Discovered." *Digital Journal*, March 24, 2015.

Szczepanski, Kallie. "How Did the Mongols Impact Europe?" ThoughtCo, February 18, 2010.

Thompson, E. P. "Time, Work-Discipline, and Industrial Capitalism." *Past & Present*, vol. 38, December 1967.

Tiflex Limited. "UK Manufacturers of Cork and Rubber Bonded Materials."

Tzu, Sun. *The Art of War: Complete Texts and Commentaries*. Shambhala, 2003.

University of Pennsylvania Museum. "Modern Mongolia: Reclaiming Genghis Khan."

Vieira, Helena. "Mechanical Clocks Prove the Importance of Technology for Economic Growth." *LSE Business Review*, September 27, 2016.

Valerio, Doug G., and Mason Mercer. "A Practical Approach to Building Isolation." [Technical Paper]. 2019.

Wayman, Erin. "Early Bow and Arrows Offer Insight Into Origins of Human Intellect." *Smithsonian Magazine*, November 7, 2012.

Whittle, Jessica. "Dissipative Seismic Design: Placing Dampers in Buildings." *The Structural Engineer*, vol. 88, no. 4, February 16, 2010.

Williams, Matt. "What Is Hooke's Law?" *Universe Today*, February 13, 2015.

Williamson, Kim. "Most Slave Shipwrecks Have Been Overlooked—Until Now." *National Geographic*, August 23, 2019.

The Worshipful Company of Clockmakers. "The Worshipful Company of Clockmakers."

Zorn, Emil A. G. Patentschrift—Erschütterungsschutz für Gebäude.pdf. 624955. [Patent filed in 1932].

Magnet

ABCMemphis. "What's inside a 113 year old hand crank telephone?" YouTube, July 26, 2020.

Akasaki, Isamu, Hiroshi Amano, and Shuji Nakamura. "Blue LEDs—Filling the World with New Light." [Nobel Prize Press Release], 2014.

Antique Telephone History. "The Gallows Telephone."

Beck, Kevin. "What Is the Purpose of a Transformer?" *Sciencing*, November 16, 2018.

Berke, Jamie. "Alexander Graham Bell and His Controversial Views on Deafness." *Verywell Health*, March 13, 2022.

Biography.com. "James West."

Bob's Old Phones. "Ericsson AC100 Series "Skeletal" Telephone."

Bondyopadhyay, Probir K. "Sir J C Bose's Diode Detector Received Marconi's First Transatlantic Wireless Signal of December 1901 (The 'Italian Navy Coherer' Scandal Revisited)." *IETE Technical Review*, vol. 15, no. 5, September 1998.

Bose, Jagadish Chandra. *Sir Jagadish Chandra Bose: His Life, Discoveries and Writings*. G. A. Natesan & Co., Madras, 1921.

Brain, Marshall. "How Radio Works." HowStuffWorks, December 7, 2000.

Brain, Marshall. "How Television Works." HowStuffWorks, November 26, 2006.

Bridge, J. A. "Sir William Brooke O'Shaughnessy, M.D., F.R.S., F.R.C.S., F.S.A.: A Biographical Appreciation by an Electrical Engineer." *Notes and Records of the Royal Society of London*, vol. 52, no. 1, 1998.

British Telephones. "Telephone No. 16."

Campbell, G. "The Evolution of Some Electromagnetic Machines." *Students' Quarterly Journal*, vol. 23, no. 89, September 1, 1952.

CERN. "Facts and Figures about the LHC."

CERN. "The Large Hadron Collider."

Chilkoti, Avantika, and Amy Kazmin. "Indian telegram service stopped stop." *Financial Times.* June 14, 2013.

"Dedication Ceremony for IEEE Milestone 'Development of Electronic Television, 1924–1941.'" IEEE Nagoya section Shizuoka University. [Paper on Ceremony].

Desai, Pratima. "Tesla's Electric Motor Shift to Spur Demand for Rare Earth Neodymium." Reuters, March 12, 2018.

Edison Tech Center. "History of Transformers."

Engineering and Technology History Wiki. "Milestones: Development of Electronic Television, 1924–1941," November 3, 2021.

Engineering and Technology History Wiki. "Milestones: First Millimeter-Wave Communication Experiments by J.C. Bose, 1894–96," November 3, 2021.

The Evolution of TV. "Kenjiro Takayanagi: The Father of Japanese Television." January 1, 2016.

Fox, Arthur. "Do Microphones Need Magnetism to Work Properly?" My New Microphone.

French, Maurice L. "Obituary: Kenjiro Takayanagi." *SMPTE Journal*, October 1990.

Garber, Megan. "India's Last Telegram Will Be Sent in July." *The Atlantic*, June 17, 2013.

Geddes, Patrick. *The Life and Work of Sir Jagadish C. Bose*. Longmans, 1920.

Ghosh, Saroj. "William O'Shaughnessy—An Innovator and Entrepreneur." *Indian Journal of History of Science*, vol. 29, no. 1, 1994.

Gorman, Mel. "Sir William O'Shaughnessy, Lord Dalhousie, and the Establishment of the Telegraph System in India." *Technology and Culture*, vol. 12, no. 4, October 1971.

Greenwald, Brian H., and John Vickrey Van Cleve. "A Deaf Variety of the Human Race": Historical Memory, Alexander Graham Bell, and Eugenics." *The Journal of the Gilded Age and Progressive Era*, vol. 14, no. 1, January 2015.

Hahn, Laura D., and Angela S. Wolters. *Women and Ideas in Engineering: Twelve Stories from Illinois*. University of Illinois Press, 2018.

Harris, Tom, Chris Pollette, and Wesley Fenlon. "How Light Emitting Diodes (LEDs) Work." HowStuffWorks, January 31, 2002.

Herbst, Jan F. "Permanent Magnets." *American Scientist*, vol. 81, no. 3, 1993.

Hillen, C. F. J. "Telephone Instruments, Payphones and Private Branch Exchanges." *Post Office Electrical Engineers Journal*, vol. 74, October 1981.

History.com. "Morse Code & the Telegraph."

History of Compass. "History of the Magnetic Compass."

Hui, Mary. "Why Rare Earth Permanent Magnets Are Vital to the Global Climate Economy." *Quartz*, May 14, 2021.

In Hamamatsu. "Takayanagi Memorial Hall."

Integrated Magnetics. "Magnets & Magnetism Frequently Asked Questions."

International Telecommunication Union. "Jagadish Chandra Bose: A Bengali Pioneer of Science."

Jessa, Tega. "Permanent Magnet." *Universe Today*, March 19, 2011.

Jha, Somesha. "RIP Telegram Fullstop." *Business Standard News*, June 15, 2013.

Joamin Gonzalez-Gutierrez. "REProMag H2020 Project." YouTube, June 10, 2016.

Kansas Historical Society. "Almon Strowger." Kansapedia.

Kantain, Tom. "Differences Between a Telephone & Telegraph." Techwalla.

Khan Academy. "Experiment: What's the Shape of a Magnetic Field?"

Kramer, John B. "The Early History of Magnetism." *Transactions of the New-comen Society*, vol. 14, no. 1, January 1, 1933.

Liffen, John. "The Introduction of the Electric Telegraph in Britain, a Reappraisal of the Work of Cooke and Wheatstone." *The International Journal for the History of Engineering & Technology*, vol. 80, no. 2, July 1, 2010.

Lincoln, Don. "How Blue LEDs Work, and Why They Deserve the Physics Nobel." Nova, October 10, 2014.

Livington, James D. *Driving Force: The Natural Magic of Magnets*. Harvard University Press, 1996.

Lucas, Jim. "What Are Radio Waves?" *Live Science*, February 27, 2019.

Magnetech, HangSeng. "Electromagnets vs Permanent Magnets." Magnets By HSMAG, May 8, 2016.

Marks, Paul. "Magnets Join Race to Replace Transistors in Computers." *New Scientist*, August 6, 2014.

Meyer, Kirstine. "Faraday and Ørsted." *Nature*, vol. 128, no. 3, August 1931.

Montgomery Ward & Co. "Rural Telephone Lines: How to Build Them." 1900 (approx.) Republished by Glen E. Razak, Kansas, 1970.

Moosa, Iessa Sabbe. "History and Development of Permanent Magnets." *International Journal for Research and Development in Technology*, vol. 2, no. 1, July 2014.

Morgan, Thaddeus. "8 Black Inventors Who Made Daily Life Easier." History.com.

MPI. "Magnetism: History of the Magnet Dates Back to 600 BC."

Naoe, Munenori. "National Institute of Information and Communications Technology." *The Journal of the Institute of Image Information and Television Engineers*, vol. 63, no. 6, 2009.

National Geographic. "Magnetism."

National Museums Scotland. "Alexander Graham Bell's Box Telephone."

NobelPrize.org. "The Nobel Prize in Physics 2014."

O'Driscoll, Bill. "Pittsburgh Author Takes a Critical Look at Alexander Graham Bell's Work with the Deaf." 90.5 WESA, April 27, 2021.

Old Telephone Books. "First Telephone Book."

O'Shaughnessy, William Brooke. *The Electric Telegraph in British India: A Manual of Instructions for the Subordinate Officers, Artificers, and Signallers Employed in the Department*. London: Printed by order of the Court of Directors, 1853.

O'Shaughnessy, William Brooke. *Memoranda Relative to Experiments on the Communication of Telegraphic Signals by Induced Electricity.* Bishop's College Press, 1839.

Overshott, K. J. "IEE Science Education & Technology Division: Chairman's Address. Magnetism: It Is Permanent." *IEE Proceedings A: Science, Measurement and Technology*, vol. 138, no. 1, 1991.

Partner, Simon. *Assembled in Japan: Electrical Goods and the Making of the Japanese Consumer.* University of California Press, 2000.

Preece, W. H., and J. Sivewright. *Telegraphy.* Ninth edition. Longmans, Green, 1891.

Pride In STEM. "Out Thinkers—Andrew Princep." YouTube, August 11, 2019.

Qualitative Reasoning Group, Northwestern University. "How Do You Make a Radio Wave?"

Queen Elizabeth Prize for Engineering. "The World's Strongest Permanent Magnet."

Ramirez, Ainissa. *The Alchemy of Us: How Humans and Matter Transformed One Another.* MIT Press, 2020.

Ramirez, Ainissa. "Jim West's Marvellous Microphone." *Chemistry World*, February 7, 2022.

Russell Kenner. "Magneto Phone Ericsson." YouTube, August 7, 2020.

The Rutland Daily Globe. "The First Newspaper Despatch [Sic.] Sent by a Humna [Sic.] Voice Over the Wires." February 12, 1877.

Salem: Still Making History. "Bell, Watson, and the First Long Distance Phone Call." March 3, 2021.

Sangwan, Satpal. "Indian Response to European Science and Technology 1757–1857." *British Journal for the History of Science*, vol. 21, 1988.

Sarkar, Suvobrata. "Technological Momentum: Bengal in the Nineteenth Century." *Indian Historical Review*, vol. 37, no. 1, June 2010.

Scholes, Sarah. "What Do Radio Waves Tell Us about the Universe?" *Frontiers for Young Minds*, February 3, 2016.

ScienceDaily. "Magnetic Fields Provide a New Way to Communicate Wirelessly: A New Technique Could Pave the Way for Ultra Low Power and High-Security Wireless Communication Systems."

Science Museum. "Goodbye to the Hello Girls: Automating the Telephone Exchange." October 22, 2018.

"Scientific Background on the Nobel Prize in Physics 2014: Efficient Blue Light-Emitting Diodes Leading to Bright and Energy-Saving White Light Sources."

Compiled by the Class for Physics of the Royal Swedish Academy of Sciences, October 7, 2014.

Sessier, Gerhard M., and James E. West. "Electrostatic Transducer." United States Patent Office 3,118,979, filed August 7, 1961, issued January 21, 1964.

Shedden, David. "Today in Media History: In 1877 Alexander Graham Bell Made the First Long-Distance Phone Call to the Boston Globe." Poynter, February 12, 2015.

Shiers, George. "Ferdinand Braun and the Cathode Ray Tube." *Scientific American*, vol. 230, no. 3, 1974.

Shridharani, Krishnalal Jethalal. *Story of the Indian Telegraphs: A Century of Progress*. Posts and Telegraph Department, 1960.

Smith, Laura. "First Commercial Telephone Exchange—Today in History: January 28." *Connecticut History*, January 28, 2020.

Smithsonian's History Explorer. "Morse Telegraph Register." November 4, 2008.

SPARK Museum of Electrical Invention. "Almon B. Strowger: The Undertaker Who Revolutionized Telephone Technology."

Strowger, A. B. "Automatic Telephone Exchange." United States Patent Office, US447918A, issued March 10, 1891.

Susmagpro. "Recovery, Reprocessing and Reuse of Rare-Earth Magnets in the Circular Economy."

Takayanagi, Kenjiro. "1926 Kenjiro Takayanagi Displays the Character on TV." [NHK Blog Post], 2002.

Technology Connections. "Lines of Light: How Analog Television Works." YouTube, July 2, 2017.

Technology Connections. "Mechanical Television: Incredibly Simple, yet Entirely Bonkers." YouTube, August 7, 2017.

Technology Connections. "Television—Playlist." YouTube.

Telephone Collectors International Inc. "TCI Library."

University of Cambridge, Department of Engineering. "Prof Hugh Hunt."

University of Oxford Department of Physics. "Cathode Ray Tube."

U.S. National Park Service. "Site of the First Telephone Exchange—National Historic Landmarks."

Vadukut, Shruti Chakraborty and Sidin. "The Telegram Is Dying." *Mint*, September 27, 2008.

Woodford, Chris. "How Does Computer Memory Work?" Explain that Stuff, July 27, 2010.

Woodford, Chris. "How Do Relays Work?" Explain that Stuff, January 4, 2009.

Woodford, Chris. "How Do Telephones Work?" Explain that Stuff, January 12, 2007.

Yanais, Hiroichi. "A Passion for Innovation—Dr. Takayanagi, a Graduate of Tokyo Tech and Pioneer of Television." Tokyo Institute of Technology.

Lens

1001 Inventions. "[FILM] 1001 Inventions and the World of Ibn Al Haytham (English Version)." YouTube, November 24, 2018.

1001 Inventions and the World of Ibn Al-Haytham. "Who Was Ibn Al-Haytham?"

Al-Amri, Mohammad D., Mohamed El-Gomati, and M. Suhail Zubairy, eds. *Optics in Our Time*. Springer International Publishing, 2016.

Aldersey-Williams, Hugh. *Dutch Light: Christiaan Huygens and the Making of Science in Europe*. Picador, 2020.

Alexander, Donavan. "Take the Perfect Shot by Understanding the Camera Lenses on Your Smartphone." *Interesting Engineering*, July 14, 2019.

Al-Khalili, Jim. "Advances in Optics in the Medieval Islamic World." *Contemporary Physics*, vol. 56, no. 2, April 3, 2015.

Al-Khalili, Jim. "Doubt Is Essential for Science—But for Politicians, It's a Sign of Weakness." *Guardian*, April 21, 2020.

Al-Khalili, Jim. "In Retrospect: Book of Optics." *Nature*, vol. 518, no. 7, 538, February 2015.

Al-Khalili, Jim. *Pathfinders: The Golden Age of Arabic Science*. Penguin Books, 2012.

UC Museum of Paleontology, University of Berkeley. "Antony van Leeuwenhoek."

Arun Murugesu, Jason. "Bionic Eye That Mimics How Pupils Respond to Light May Improve Vision." *New Scientist*, March 17, 2022.

Ball, Philip. "Ibn Al Haytham and How We See." *Science Stories*, BBC, January 9, 2019.

Beller, Jonathan. *The Message Is Murder: Substrates of a Computational Capital*. Pluto Press, 2017.

Botchway, Stanley W., P. Reynolds, A. W. Parker, and P. O'Neill. "Use of Near Infrared Femtosecond Lasers as Sub-Micron Radiation Microbeam for Cell DNA Damage and Repair Studies." *Mutation Research*, vol. 704, 2010.

Botchway, Stanley W., Kathrin M. Scherer, Steve Hook, Christopher D. Stubbs,

Eleanor Weston, Roger H. Bisby, and Anthony W. Parker. "A Series of Flexible Design Adaptations to the Nikon E-C1 and E-C2 Confocal Microscope Systems for UV, Multiphoton and FLIM Imaging: NIKON CONFOCAL FOR UV MULTIPHOTON AND FLIM." *Journal of Microscopy*, vol. 258, no. 1, April 2015.

Branch Education. "What's Inside a Smartphone?" YouTube, July 11, 2019.

BrianJFord.com. "Brian J Ford's 'Leeuwenhoek Legacy .'"

The British Museum. "Inlay | British Museum (Nimrud)."

California Center for Reproductive Medicine—CACRM. "Understanding Embryo Grading & Blastocyst Grades | What Do Embryo Grades Mean? CACRM." YouTube, June 13, 2014.

Carrington, David. "How Many Photos Will Be Taken in 2020?" *Mylio*, April 29, 2021.

Cobb, M. "An Amazing 10 Years: The Discovery of Egg and Sperm in the 17th Century: The Discovery of Egg and Sperm." *Reproduction in Domestic Animals*, vol. 47, August 2012.

Cole, Teju. "When the Camera Was a Weapon of Imperialism. (And When It Still Is.)." *New York Times*, February 6, 2019.

Cooper Surgical. "Equipment: Our Cutting-Edge Range for ART—Incubators, Workstations, Micromanipulators and Lasers." [Technical Brochure].

CooperSurgical Fertility Companies. "RI Integra 3." September 27, 2019.

Cox, Spencer. "What Is F-Stop, How It Works and How to Use It in Photography." *Photography Life*, January 6, 2017.

Deol, Simar. "Remembering Homai Vyarawalla, India's First Female Photojournalist." *INDIE Magazine*, March 12, 2020.

Digital Public Library of America. "Early Photography."

The Economist. "Taking Selfies with a Liquid Lens." April 14, 2021.

Fermilab. "Why Does Light Bend When It Enters Glass?" YouTube, May 1, 2019.

Fertility Associated. "ICSI Footage." YouTube, March 13, 2017.

Fertility Specialist Sydney. "Ivf Embryo Developing Over 5 Days by Fertility Dr. Raewyn Teirney." YouTube, April 12, 2014.

Fineman, Mia. "Kodak and the Rise of Amateur Photography." The Metropolitan Museum of Art: Heilbrunn Timeline of Art History. October 2004.

Ford, Brian J. "Celebrating Leeuwenhoek's 375th Birthday: What Could His Microscopes Reveal?" *Infocus Magazine*, December 2007.

Ford, Brian J. "The Cheat and the Microscope: Plagiarism Over the Centuries." *The Microscope*, vol. 53, no. 1, 2010.

Ford, Brian J. "Found: The Lost Treasure of Anton van Leeuwenhoek." *Science Digest*, vol. 90, no. 3, March 1982.

Ford, Brian J. *The Optical Microscope Manual: Past and Present Uses and Techniques.* David & Charles (Holdings) Limited, 1973.

Ford, Brian J. "Recording Three Leeuwenhoek Microscopes." *Infocus Magazine*, December 6, 2015.

Ford, Brian J. "The Royal Society and the Microscope." *Notes and Records of the Royal Society of London*, vol. 55, no. 1, January 22, 2001.

Ford, Brian J. *Single Lens: The Story of the Simple Microscope.* Harper & Row, 1985.

Ford, Brian J. "The Van Leeuwenhoek Specimens." *Notes and Records of the Royal Society of London*, vol. 36, no. 1, August 1981.

Gates, Henry Louis, Jr. "Frederick Douglass's Camera Obscura: Representing the Antislave 'Clothed and in Their Own Form.'" *Critical Enquiry*, vol. 42, Autumn 2015.

Gauweiler, Lena, Dr. Eckhardt, and Dr. Behler. "Optische Pinzette (optical tweezer)." Presented at the Laseranwendungstechnik WS 19/20 December 17, 2019.

Gest, H. "The Discovery of Microorganisms by Robert Hooke and Antoni van Leeuwenhoek, Fellows of The Royal Society." *Notes and Records of the Royal Society of London*, vol. 58, no. 2, May 22, 2004.

Gregory, Andrew. "Bionic Eye Implant Enables Blind UK Woman to Detect Visual Signals." *Guardian*, January 21, 2022.

Gross, Rachel E. "The Female Scientist Who Changed Human Fertility Forever." BBC.

Hall, A. R. "The Leeuwenhoek Lecture, 1988, Antoni Van Leeuwenhoek 1632–1723." *Notes and Records of the Royal Society of London*, vol. 43, 1989.

Hand, Eric. "We Need a People's Cryo-EM." Scientists Hope to Bring Revolutionary Microscope to the Masses." *Science*, January 23, 2020.

Hannavy, J, ed. "LENSES: 1830s–1850s." In *Encyclopedia of Nineteenth Century Photography.* London: Routledge, 2008.

Haque, Nadeem. "Author Bradley Steffens on 'First Scientist,' Ibn al- Haytham." *Muslim Heritage*, January 8, 2020.

Helff, Sissy, and Stefanie Michels. "Chapter: Re-Framing Photography—Some

Thoughts." In *Global Photographies: Memory, History, Archives*, Transcript Verlag, 2021.

Hertwig, Oskar. *Dokutmente Zur Geschichte Der Zeugungslehre: Eine Historische Studie*. Verlag von Friedrich Cohen, 1918.

History of Science Museum. "Sphere No. 8: Thomas Sutton Panoramic Camera Lens." Autumn 1998.

IIT Bombay July 2018. "Week 5-Lecture 27: Ti:Sapphire Laser (Lab Visit)." You-Tube, February 20, 2020.

Jain, Mahima. "The Exoticised Images of India by Western Photographers Have Left a Dark Legacy." Scroll.in, February 20, 2019.

Koenen, Anke, and Michael Zolffel. *Microscopy for Dummies*. Zeiss, 2020.

Kress, Holger. *Cell Mechanics During Phagocytosis Studied by Optical Tweezers Based Microscopy*. Cuvillier Verlag, 2006.

Kriss, Timothy C., and Vesna Martich Kriss. "History of the Operating Microscope: From Magnifying Glass to Microneurosurgery." *Neurosurgery*, vol. 42, no. 4, 1998.

Kuo, Scot C. "Using Optics to Measure Biological Forces and Mechanics." *Traffic*, vol. 2, no. 11, 2001.

Lawrence, Iszi. "Animalcules." *The Z-List Dead List*, season 3, episode 3, February 26, 2015.

Leica Microsystems. "Leica Objectives: Superior Optics for Confocal and Multiphoton Research Microscopy." [Technical Brochure], 2014.

Leica Microsystems. "Leica TCS SP8 STED: Opening the Gate to Super-Resolution." [Technical Brochure], 2012.

Leica Microsystems. "Leica TCS SP8 STED 3X: Your Next Dimension!" [Technical Brochure], 2014.

"Lens History." In *The Focal Encyclopedia of Photography*, Desk edition. London: Focal Press, 2017.

Lens on Leeuwenhoek. "Specimens: Sperm."

Lerner, Eric J. "Advanced Applications: Biomedical Lasers: Lasers Support Biomedical Diagnostics." *Laser Focus World*, May 1, 2000.

Maison Nicéphore Niépce. "Niépce and the Invention of Photography."

Marsh, Margaret, and Wanda Ronner. *The Pursuit of Parenthood: Reproductive Technology from Test-Tube Babies to Uterus Transplants*. Johns Hopkins University Press, 2019.

McConnell, Anita. *A Survey of the Networks Bringing a Knowledge of Optical Glass-Working to the London Trade, 1500–1800.* Cambridge: Whipple Museum of the History of Science, 2016.

McQuaid, Robert. "Ibn Al-Haytham, the Arab Who Brought Greek Optics into Focus for Latin Europe." *MedCrave Online*, April 12, 2019.

Medline Plus. "Laser Therapy."

The Metropolitan Museum of Art. "Collection Item: Unknown | [Amateur Snapshot Album]."

Microscope World. "ZEISS Axio Observer Inverted Life Sciences Research Microscope."

Mokobi, Faith. "Inverted Microscope—Definition, Principle, Parts, Labeled Diagram, Uses, Worksheet." *Microbe Notes*, April 10, 2022.

Mourou, Gérard, and Donna Strickland. "Tools Made of Light." The Nobel Prize in Physics 2018: Popular Science Background. The Royal Swedish Academy of Sciences.

Narayan, Roopa H. "Nyaya-Vaisheshika: The Theory of Matter in Indian Physics."

National Science and Media Museum. "The History of Photography in Pictures." March 8, 2017.

NewsCenter. "Chirped-Pulse Amplification: 5 Applications for a Nobel Prize–Winning Invention," October 4, 2018.

Nield, David. "The Extra Lenses in Your Smartphone's Camera, Explained." *Popular Science*, March 28, 2019.

Nikon. "The Optimal Parameters for ICSI—Perfect Your ICSI with Precise Optics." [Information Brochure], 2019.

W. W. Norton & Company. "Picturing Frederick Douglass."

Open University. "Life through a Lens." March 2, 2020.

Pearey Lal Bhawan. "How the Invention of Photography Changed Art."

Photo H26. "Périscope Apple : ceci n'est pas un zoom." April 22, 2016.

Photonics. "Lasers: Understanding the Basics."

Pool, Rebecca. "Life through a Microscope: Profile—Professor Brian J Ford." *Microscopy and Analysis*, October 2017.

Poppick, Laura. "The Long, Winding Tale of Sperm Science." *Smithsonian Magazine*, June 7, 2017.

Powell, Martin. *Louise Brown: 40 Years of IVF, My Life as the World's First Test-Tube Baby.* Bristol Books, 2018.

Pritchard, Michael. *A History of Photography in 50 Cameras.* Bloomsbury, 2019.

Randomtronic. "Close Look at Mobile Phone Camera Optics." YouTube, December 10, 2016.

Rehm, Lars. "Ultra-Thin Lenses Could Eliminate the Need for Smartphone Camera Bumps." *DPReview*, October 12, 2019.

Rock, John, and Miriam F. Menkin, "In Vitro Fertilization and Cleavage of Human Ovarian Eggs," *Science, New Series*, vol. 100, issue 2588, August 4, 1944, 105–7.

The Royal Society. "Arabick Roots." June 2011.

The Royal Society. "Eye to Eye with a 350-Year Old Cow: Leeuwenhoek's Specimens and Original Microscope Reunited in Historic Photoshoot."

Scheisser, Tim. "Know Your Smartphone: A Guide to Camera Hardware." *TechSpot*, July 28, 2014.

Sines, George, and Yannis A. Sakellarakis. "Lenses in Antiquity." *American Journal of Archaeology*, vol. 91, no. 2., 1987.

Stierwalt, Sabrina. "A Nobel Prize–Worthy Idea: What Is Chirped Pulse Amplification?" Quick and Dirty Tips, February 12, 2019.

Subcon Laser Cutting Ltd. "Contributions of Laser Technology to Society." September 24, 2019.

Szczepanski, Kallie. "Kites, Maps, Glass and Other Asian Inventions." ThoughtCo, December 13, 2019.

Tbakhi, Abdelghani, and Samir S. Amr. "Ibn Al-Haytham: Father of Modern Optics." *Annals of Saudi Medicine*, vol. 27, no. 6, 2007.

van Leeuwenhoek, Antoni. "Leeuwenhoek's Letter to the Royal Society (Dutch)." Circulation of Knowledge and Learned Practices in the 17th Century Dutch Republic.

van Mameren, Joost. "Optical Tweezers: Where Physics Meets Biology." *Physics World*, November 13, 2008.

Wheat, Stacy, Katie Vaughan, and Stephen James Harbottle. "Can Temperature Stability Be Improved During Micromanipulation Procedures by Introducing a Novel Air Warming System?" *Reproductive BioMedicine Online*, vol. 28, May 2014.

Wired. "Photography Snapshot: The Power of Lenses." September 14, 2012.

Woodford, Chris. "How Do Lasers Work? Who Invented the Laser?" Explain that Stuff, April 8, 2006.

Zeiss. "Assisted Reproductive Technology." [Technical Brochure 2.0].

String

Arie, Purushu. "Caste, Clothing and the Bias Cut." *The Voice of Fashion*, June 7, 2021.

Astbury, W. T., and A. Street. "X-Ray Studies of the Structure of Hair, Wool, and Related Fibres. I. General." *Philosophical Transactions of the Royal Society of London: Series A, Containing Papers of a Mathematical or Physical Character*, vol. 230, 1932.

BBC News. "50,000-Year-Old String Found at France Neanderthal Site," April 13, 2020.

Bellis, Mary. "Information About Textile Machinery Inventions." ThoughtCo, July 1, 2019.

Bilal, Khadija. "Here's Why It All Changed: Pink Used to Be a Boy's Color & Blue for Girls." *The Vintage News*, May 1, 2019.

Brown, Theodore M., and Elizabeth Fee. "Spinning for India's Independence." *American Journal of Public Health*, vol. 98, no. 1, January 2008.

Castilho, Cintia J., Dong Li, Muchun Liu, Yue Liu, Huajian Gao, and Robert H. Hurt. "Mosquito Bite Prevention Through Graphene Barrier Layers." *Proceedings of the National Academy of Sciences*, vol. 116, no. 37, September 10, 2019.

Chen, Cathleen. "Why Genderless Fashion Is the Future." *The Business of Fashion*, November 22, 2019.

Clase, Catherine, Charles-Francois de Lannoy, and Scott Laengert. "Polypropylene, the Material Now Recommended for COVID-19 Mask Filters: What It Is, Where to Get It." *The Conversation*, November 19, 2020.

Edden, Shetara. "High-Tech Performance Fabrics to Know." *Maker's Row*, October 12, 2016.

Firth, Ian P. T., and Poul Ove Jensen. "Bridges: Spanning Art and Technology." *The Structural Engineer*, Centenary Issue, July 21, 2008.

Freyssinet. "H 1000 Stay Cable System." 2014.

Gersten, Jennifer. "Are Catgut Instrument Strings Really Made from Cat Guts? The Answer Might Surprise You." WQXR, July 17, 2017.

Gruen, L. C., and E. F. Woods. "Structural Studies on the Microfibrillar Proteins of Wool." *Biochemical Journal*, vol. 209, 1983.

Hardy, B. L., M. H. Moncel, C. Kerfant, M. Lebon, L. Bellot-Gurlet, and N. Mélard. "Direct Evidence of Neanderthal Fiber Technology and Its Cognitive and Behavioral Implications." *Scientific Reports*, vol. 10, no. 1, December 2020.

Hagley Magazine. "Stephanie Kwolek Collection Arrives." *Hagley Magazine*, Winter 2014.

History of Clothing. "History of Clothing—History of Fabrics and Textiles."

Hock, Charles W. "Structure of the Wool Fiber as Revealed by the Microscope." *The Scientific Monthly*, vol. 55, no. 6, December 1942.

Huang, Belinda. "What Kind of Impact Does Our Music Really Make on Society?" *Sonic Bids*, August 24, 2015.

Hudson-Miles, Richard. "New V&A Menswear Exhibition: Fashion Has Always Been at the Heart of Gender Politics." *The Conversation*, March 24, 2022.

India Instruments. "Tanpura."

Jabbr, Ferris. "The Long, Knotty, World-Spanning Story of String." *Hakai Magazine*, March 6, 2018.

Jones, Lucy. "Six Fashion Materials That Could Help Save the Planet." BBC Earth.

Kakodkar, Priyanka. "Miraj's Legacy Sitar-Makers Go Online to Survive." *Times of India*, July 15, 2018.

Kittler, Ralf, Manfred Kayser, and Mark Stoneking. "Molecular Evolution of Pediculus Humanus and the Origin of Clothing." *Current Biology*, vol. 13, August 19, 2003.

Kwolek, Stephanie Louise. "Optimally Anisotropic Aromatic Polyamide Dopes." United States Patent Office, 3,671,542, filed May 23, 1969, issued June 20, 1972.

Lim, Taehwan, Huanan Zhang, and Sohee Lee. "Gold and Silver Nanocomposite-Based Biostable and Biocompatible Electronic Textile for Wearable Electromyographic Biosensors." *APL Materials*, vol. 9, no. 9, September 1, 2021.

Macalloy. "McCalls Special Products Ltd—Historical Background." [Company Brochure], August 7, 2002.

Mansour, Katerina. "Sustainable Fashion Finds Success in New Materials." *Early Metrics*, April 15, 2021.

Marcal, Katrine. *Mother of Invention: How Good Ideas Get Ignored in an Economy Built for Men.* William Collins, 2021.

McCullough, David. *The Great Bridge: The Epic Story of the Building of the Brooklyn Bridge.* Simon & Schuster Paperbacks, 1972.

McFadden, Christopher. "Mechanical Engineering in the Middle Ages: The Catapult, Mechanical Clocks and Many More We Never Knew About." *Interesting Engineering*, April 28, 2018.

Museum of Design Excellence. "Charkha, the Device That Charged India's Freedom Movement." Google Arts & Culture.

Myerscough, Matthew. "Suspension Bridges: Past and Present." *The Structural Engineer*, vol. 10, July 2013.

New World Encyclopedia. "String Instrument."

New World Encyclopedia. "Textile Manufacturing."

Nuwer, Rachel. "Lice Evolution Tracks the Invention of Clothes." *Smithsonian Magazine*, November 14, 2012.

Okie, Suz. "These Materials Are Replacing Animal-Based Products in the Fashion Industry." *World Economic Forum*, October 6, 2021.

Plata, Allie. "Q'eswachaka, the Last Inka Suspension Bridge." *Smithsonian* Magazine, August 4, 2017.

Ploszajski, Anna. *Handmade: A Scientist's Search for Meaning Through Making.* Bloomsbury, 2021.

Postrel, Virginia. "How Job-Killing Technologies Liberated Women." Bloomberg, March 14, 2021.

Raman, C. V. "On Some Indian Stringed Instruments." *Indian Association for the Cultivation of Science*, vol. 7, 1921.

Ramirez, Catherine S. *The Woman in the Zoot Suit: Culture, Nationalism and the Politics of Memory.* Duke University Press, 2009.

Raniwala, Praachi. "India's Long History with Genderless Clothing." *Mint Lounge*, December 16, 2020.

Reuters. "Bridge Made of String: Peruvians Weave 500-Year-Old Incan Crossing Back into Place." *Guardian*, June 16, 2021.

Rippon, J. A. "Wool Dyeing." In *The Structure of Wool*. Bradford (UK): Society of Dyers and Colourists, 1992.

Roda, Allen. "Musical Instruments of the Indian Subcontinent." The Metropolitan Museum of Art: Heilbrunn Timeline of Art History, March 2009.

Sears, Clare. *Arresting Dress: Cross-Dressing, Law, and Fascination in Nineteenth-Century San Francisco.* Duke University Press, 2015.

Sewell, Abby. "Photos of the Last Incan Suspension Bridge in Peru." *National Geographic*, August 31, 2018a.

Sievers, Christine, Lucinda Backwell, Francesco d'Errico, and Lyn Wadley. "Plant Bedding Construction between 60,000 and 40,000 Years Ago at Border Cave, South Africa." *Quaternary Science Reviews*, vol. 275, January 2022.

Skope. "A Brief History of String Instruments." May 6, 2013.

Steel Wire Rope. "All Wire Ropes."

String Ovation Team. "How Are Violin Strings Made?" *Connolly Music*, March 7, 2019.

SWR. "Sourcing, Designing and Producing Wire Rope Solutions." [Company Brochure].

Talati-Parikh, Sitanshi. "Why Are School Uniforms Still Gendered?" *The Swaddle*, May 13, 2018.

Talbot, Jim. "First Steel-Wire Suspension Bridge." *Modern Steel Construction*, June 2011.

Tecni Ltd. "Low Rotation Wire Rope—19 x 7 Construction Cable." YouTube, July 18, 2019.

Toss Levy. "Tanpura History."

Toss Levy, Indian Musical Instruments. "The Correct Use of the Tanpura Jiva (Threads)." YouTube, August 3, 2020.

UNESCO. "Did You Know? The Exchange of Silk, Cotton and Woolen Goods, and Their Association with Different Modes of Living Along the Silk Roads."

Urmi Battu. "How to Tune a Tanpura." YouTube, March 16, 2021.

Vaid-Menon, Alok. *Beyond the Gender Binary*. Penguin Workshop, 2020.

Venkataraman, Vaishnavi. "Soon, You Can Zip-Line from Ferrari World Abu Dhabi's Stunning Roof." *Curly Tales*, October 22, 2020.

Vincent, Susan J. *The Anatomy of Fashion*. Berg, 2009.

Walstijn, Maarten van, Jamie Bridges, and Sandor Mehes. "A Real-Time Synthesis Oriented Tanpura Model." In *Proceedings of the 19th International Conference on Digital Audio Effects (DAFx-16)*. Brno, 2016.

Whitfield, John. "Lice Genes Date First Human Clothes." *Nature*, August 20, 2003.

Willson, Tayler. "Meet the Emerging Brand Making Sneakers from Coffee Grounds." *Hypebeast*, August 12, 2021.

"Wool: Raw Wool Specification." *Encyclopedia of Polymer Science and Technology, Wood Composites*, vol. 12.

World Health Organization. "Coronavirus Disease (COVID-19): Masks." January 5, 2022.

Wragg Sykes, Rebecca. *Kindred: Neanderthal Life, Love, Death and Art*. Bloomsbury, 2020.

Pump

1001 Inventions. "5 Amazing Mechanical Devices from Muslim Civilization."

Abbott. "About the HeartMate II LVAD."

Abbott. "HeartMate 3 LVAD."

Abbott. "How the CentriMag Acute Circulatory Support System Works."

Al-Hassani, Salim. "Al-Jazari: The Mechanical Genius." *Muslim Heritage*, February 9, 2001.

Al-Hassani, Salim. "The Machines of Al-Jazari and Taqi Al-Din." *Muslim Heritage*, December 30, 2004.

Al-Hassani, Salim. "Al-Jazari's Third Water-Raising Device: Analysis of Its Mathematical and Mechanical Principles." *Muslim Heritage*, April 24, 2008.

Ameda. "Our History."

Anderson, Brooke, J. Nealy, Garry Qualls, Peter Staritz, John Wilson, M. Kim, Francis Cucinotta, William Atwell, G. DeAngelis, and J. Ware. "Shuttle Spacesuit (Radiation) Model Development." *SAE Technical Papers*, February 1, 2001.

Bazelon, Emily. "Milk Me: Is the Breast Pump the New BlackBerry?" *Slate*, March 27, 2006.

Behe, Caroline. "Transgender & Non-Binary Parents." La Leche League International.

bigclivedotcom. "Inside a Near-Silent Piezoelectric Air Pump." YouTube, June 14, 2018.

B. L. S., Amrit. "Why the US Pig Heart Transplant Was Different from the 1997 Assam Doc's Surgery." *The Wire Science*, January 13, 2022.

Bologna, Caroline. "200 Years of Breast Pumps, in 18 Images." HuffPost UK, August 1, 2016a.

British Heart Foundation. "Focus on: Left Ventricular Assist Devices."

British Heart Foundation. "How Your Heart Works."

Butler, Karen. "Relactation and Induced Lactation." *La Leche League GB*, March 19, 2016.

Cadogan, David. "The Past and Future Space Suit." *American Scientist*, vol. 103, no. 5, 2015.

Campbell, Dallas. *Ad Astra: An Illustrated Guide to Leaving the Planet.* Simon & Schuster, 2017.

CBS News. "The Seamstresses Who Helped Put a Man on the Moon." July 14, 2019.

Cheng, Allen, Christine A. Williamitis, and Mark S. Slaughter. "Comparison of Continuous-Flow and Pulsatile-Flow Left Ventricular Assist Devices: Is

There an Advantage to Pulsatility?" *Annals of Cardiothoracic Surgery*, vol. 3, no. 6, November 2014.

Chu, Jennifer. "Shrink-Wrapping Spacesuits." Massachusetts Institute of Technology, September 18, 2014.

Davis, Charles Patrick. "How the Heart Works: Diagram, Anatomy, Blood Flow." MedicineNet.

Diana West. "Trans Breastfeeding FAQ."

Dinerstein, Joel. "Technology and Its Discontents: On the Verge of the Posthuman." *American Quarterly*, vol. 58, no. 3, 2006.

Elvie. "Elvie."

Encyclopedia of Australian Science and Innovation. "Robinson, David – Person." Swinburne University of Technology, Centre for Transformative Innovation.

Encyclopedia Britannica. "Shaduf: Irrigation Device."

The European Space Agency. "Alexei Leonov: The Artistic Spaceman." October 4, 2007.

Eurostemcell. "The Heart: Our First Organ."

Garber, Megan. "A Brief History of Breast Pumps." *The Atlantic*, October 21, 2013.

Greatrex, Nicholas, Matthias Kleinheyer, Frank Nestler, and Daniel Timms. "This Maglev Heart Could Keep Cardiac Patients Alive." *IEEE Spectrum*, August 22, 2019.

Greenfield, Rebecca. "Celebrity Invention: Paul Winchell's Artificial Heart." *The Atlantic*, January 7, 2011.

Hamzelou, Jessica. "Transgender Woman Is First to Be Able to Breastfeed Her Baby." *New Scientist*, February 14, 2018.

Hasic, Albinko. "The First Spacewalk Could Have Ended in Tragedy for Alexei Leonov. Here's What Went Wrong." *Time*, March 18, 2020.

History.com. "March 23: Artificial Heart Patient Dies."

How Products Are Made. "Spacesuit."

Jarvik Heart. "Robert Jarvik, MD on the Jarvik-7." April 6, 2016.

Kato, Tomoko S., Aalap Chokshi, Parvati Singh, Tuba Khawaja, Faisal Cheema, Hirokazu Akashi, Khurram Shahzad, et al. "Effects of Continuous-Flow Versus Pulsatile-Flow Left Ventricular Assist Devices on Myocardial Unloading and Remodeling." *Circulation: Heart Failure*, vol. 4, no. 5, September 2011.

Kotz, Deborah. "2022 News—University of Maryland School of Medicine Faculty Scientists and Clinicians Perform Historic First Successful Transplant of

Porcine Heart into Adult Human with End-Stage Heart Disease." University of Maryland School of Medicine, January 10, 2022.

Kwan, Jacklin. "What Would Happen to the Human Body in the Vacuum of Space?" *Live Science*, November 13, 2021.

Lathers, Marie. *Space Oddities: Women and Outer Space in Popular Film and Culture, 1960–2000*. Bloomsbury Publishing, 2010.

Ledford, Heidi. "Ghost Heart Has a Tiny Beat." *Nature*, January 13, 2008.

Le Fanu, James. *The Rise and Fall of Modern Medicine*. Abacus, 2011.

Longmore, Donald. *Spare Part Surgery: The Surgical Practice of the Future*. Aldus Books London, 1968.

Madrigal, Alexis C. "The World's First Artificial Heart." *The Atlantic*, October 1, 2010.

Mahoney, Erin. "Spacesuit Basics." NASA, October 4, 2019.

Martucci, Jessica. "Breast Pumping." *AMA Journal of Ethics*, vol. 15, no. 9, September 1, 2013.

McFadden, Christopher. "Mechanical Engineering in the Middle Ages: The Catapult, Mechanical Clocks and Many More We Never Knew About." *Interesting Engineering*, April 28, 2018.

McKellar, Shelley. *Artificial Hearts: The Allure and Ambivalence of a Controversial Medical Technology*. Wellcome Collection, 2018.

Mechanical Boost. "What Is a Pump? What Are the Types of Pumps?" December 4, 2020.

MedicineNet. "Picture of Heart Detail."

Medlife Crisis. "The 6 Weirdest Hearts in the Animal Kingdom." YouTube, February 11, 2018.

Mends, Francine. "What Are Piezoelectric Materials?" *Sciencing*, December 28, 2020.

Morris, Thomas. *The Matter of the Heart: A History of the Heart in Eleven Operations*. Vintage, 2017.

Mullin, Emily. "A Simple Artificial Heart Could Permanently Replace a Failing Human One." *MIT Technology Review*, March 16, 2018.

Murata Manufacturing Co. Ltd. "Basic Knowledge of Microblower (Air Pump)."

Murata Manufacturing Co. Ltd. "Microblower (Air Pump) | Micro Mechatronics."

National Heart, Lung and Blood Institute. "Developing a Bio-Artificial Heart."

National Heart, Lung and Blood Institute. "What Is Total Artificial Heart?"

National Museum of American History. "Liotta-Cooley Artificial Heart."

Newman, Dava. "Building the Future Spacesuit." *ASK Magazine*.

O'Donahue, Kelvin. "How Do Oil Field Pumps Work?" Sciencing, March 14, 2018.

Pumps and Systems. "History of Pumps." February 28, 2018.

Sarkar, Manjula, and Vishal Prabhu. "Basics of Cardiopulmonary Bypass." *Indian Journal of Anaesthesia*, vol. 61, no. 9, September 2017.

Science Friday. "Bringing a 'Ghost Heart' to Life." February 14, 2020.

Science Museum Group. "Sir Henry Wellcome's Museum Collection."

Shrouk El-Attar (@dancingqueerofficial). "Chatting with @elvie's CEO and MY BOSS @tania.Boler." Instagram, March 11, 2021.

Shumacker, Harris B. *A Dream of the Heart: The Life of John H. Gibbon, Jr., Father of the Heart-Lung Machine*, 1999.

Smithsonian Magazine. "The Nightmare of Voskhod 2." *Smithsonian Magazine*, January 2005.

The Stemettes Zine. "Meet Vinita Marwaha Madill." January 11, 2021.

SynCardia. "7 Things You Should Know About Artificial Hearts," August 9, 2018.

SynCardia. "SynCardia Temporary Total Artificial Heart."

Taschetta-Millane, Melinda. "Pig Heart Transplant Patient Continues to Thrive." DAIC, February 16, 2022.

TED Archive. "How to Create a Space Suit—Dava Newman." YouTube, August 29, 2017.

Texas Heart Institute. "50th Anniversary of the World's First Total Artificial Heart."

Thomas, Kenneth S. "The Apollo Portable Life Support System." NASA.

Thomas, Kenneth S., and Harold J. McMann. *U.S. Spacesuits*. Second Edition. Springer-Praxis, 2012.

Thornton, Mike, Dr. Robert Randall, and Kurt Albaugh. "Then and Now: Atmospheric Diving Suits." *Underwater Magazine*, March/April 2001.

US Patents Office. "Breast Pump System Patent Application—USPTO report."

Vallely, Paul. "How Islamic Inventors Changed the World." *Independent*, May 17, 2008.

VanHemert, Kyle. "Aerospace Gurus Show Off a Fancy Space Suit Made for Mars." *Wired*, November 5. 2014.

Watts, Sarah. "The Voice Behind Some of Your Favorite Cartoon Characters Helped Create the Artificial Heart." Leaps.org, July 30, 2021.

WebMD. "Anatomy and Circulation of the Heart."

Wellcome Collection. "A Breast Pump Manufactured by H. Wright. Wood."

Winderlich, Melanie. "How Breast Pumps Work." How Stuff Works, February 9, 2012.

World Pumps. "A Brief History of Pumps." March 6, 2014.

INDEX

Page numbers in *italics* refer to illustrations.